The Chemistry of Inkjet Inks

The Chemistry of Inkjet Inks

Editor

Shlomo Magdassi

The Hebrew University of Jerusalem, Israel

World Scientific

NEW JERSEY · LONDON · SINGAPORE · BEIJING · SHANGHAI · HONG KONG · TAIPEI · CHENNAI

Published by

World Scientific Publishing Co. Pte. Ltd.

5 Toh Tuck Link, Singapore 596224

USA office: 27 Warren Street, Suite 401-402, Hackensack, NJ 07601

UK office: 57 Shelton Street, Covent Garden, London WC2H 9HE

Library of Congress Cataloging-in-Publication Data
The chemistry of inkjet inks / editor, Shlomo Magdassi.
    p. cm.
   Includes bibliographical references and index.
   ISBN-13 978-981-281-821-8
   ISBN-10 981-281-821-9
  1. Printing ink. 2. Ink-jet printing. I. Magdassi, Shlomo, 1954–

2010287021

British Library Cataloguing-in-Publication Data
A catalogue record for this book is available from the British Library.

First published 2010 (Hardcover)
Reprinted 2016 (in paperback edition)
ISBN 978-981-3203-49-5

Typeset by Stallion Press
Email: enquiries@stallionpress.com

Printed in Singapore

Contents

Preface

Modern printing is based on digitizing information, and representation of the information on a substrate, such as paper, pixel by pixel. One of the most abundant methods of digital printing is through inkjet printers. These printers are widely used in office and home, and in industrial applications such as wide format printing. Until recently, most inkjet printing was performed in graphic applications, i.e., converting conventional printing of documents into digital printing. Inkjet printing was found to be so powerful, that the method was adopted to print various functional materials, such as conductive inks, light emitting diodes (LEDs), and even three dimensional structures. A reflection of this very active field is the large number of scientific and industrial conferences which takes place every year, and the huge number of patents which are published each year. Recently, there appears to be an increasing number of scientific papers on this subject, mainly focused on printing functional materials and unique properties of the printed patterns.

The inkjet printing process is very complicated, and requires delicate tailoring of the chemical and physicochemical properties of the ink. The ink should meet the requirements which are related to storage stability, jetting performance, color management (in the case of graphic printing), wetting and adhesion on substrates. Obviously, these requirements, which represent different scientific disciplines, such as colloid chemistry, physics and chemical engineering, indicate the need for an interdisciplinary book, which will cover all aspects of making and utilizing inkjet inks.

As can be seen in the table of content, the book provides basic and essential information on the important parameters which determine the ink performance, on ink formulations, and also provides insight into novel and exciting applications based on inkjet printing of functional materials. Therefore, I hope that the book will serve the large community of industrial chemists who deal with ink formulations and synthesis of chemicals for inks, chemical engineers and physicists which deal with rheological and flow properties of inks, as well as scientists in academic institutes who seek to develop novel applications based on inkjet printing of new materials. The various chapters of the book are written by experts from academic institutions as well as from leading companies in the field of ink formulations and raw materials manufacturing.

The first five chapters of the book focus on fundamental aspects of printing technologies, pigments and ink formulations and, and interactions of the inks with the substrates. The next six chapters focus on actual inkjet inks formulations and raw materials, by discussing the main groups of inks: waterborne, solvent-based, and UV inks. The last five chapters present unique ink systems and functional inks, such as those for obtaining 3D structures or printed electronic devices.

I would like to thank all the authors who put so much efforts to enable the publishing of this book. I also thank Dr. Vinetsky for her great help in finalizing the book, and the very professional team of World Scientific Publishing Co. Last but not least, many thanks to all my students who are performing exciting research on new materials and novel applications of inkjet printing.

Professor Shlomo Magdassi
The Hebrew University of Jerusalem, Israel
February 2009, Jerusalem

PART I

BASIC CONCEPTS

CHAPTER 1

Inkjet Printing Technologies

Alan Hudd
Xennia Technology Limited

INTRODUCTION

Inkjet has become a household word through its ubiquitous presence on the consumer desktop as a low cost, reliable, quick, and convenient method of printing digital files. Although inkjet technology has been utilized since the 1950s in products such as medical strip chart recorders by Siemens,[1] and has seen commercial success in high speed date coding equipment since the 1970s,[2] the potential impact of the technology in industrial applications is only now becoming widely recognized.

In theory, inkjet is simple. A print head ejects tiny drops of ink onto a substrate. In practice, implementation of the technology is complex and requires multidisciplinary skills. Reliable operation depends on careful design, implementation, and operation of a complete system where no element is trivial.

Given the underlying complexity, what drives the industrial adoption of inkjet? The characteristics of inkjet technology offer advantages to a wide range of applications. Inkjet is increasingly viewed as more than just a printing or marking technique. It can also be used to apply coatings, to accurately deposit precise amounts of materials, and even to build micro or macro structures. The list of industrial uses for inkjet technology seems endless and includes the reduction of manufacturing costs, provision of higher quality output, conversion of processes from analogue to digital, reduction in inventory, the new ability to process larger, smaller, or more flexible, fragile, or non-flat substrates, reduction of waste, mass customization, faster prototyping, and implementation of just-in-time manufacturing.

The introduction of industrial inkjet technology into manufacturing environments can provide a modest improvement, or it can prove to be revolutionary; the commercial benefits are usually obvious.

CURRENT AND EMERGING MARKETS

Commercially successful implementations of industrial inkjet technology include high speed coding or marking of packages or products, mail addressing, the manufacture of simulated-wood doors and furniture, and wide format graphics for indoor and outdoor signs and posters, trade show displays, billboards, and banners.

Emerging applications range from utilitarian to glamorous. Up and coming industrial applications include the decoration of textiles, ceramics, and food; using inkjet to replace existing analogue manufacturing processes such as pad printing, screen printing, spraying, roll coating, and dipping; and the introduction of high speed digital narrow web presses to enhance (or in some cases replace) analogue high speed flexographic or offset lithographic printing equipment for applications like labels, magazines, or books on demand. Particularly hot topics that receive a great deal of press attention and research focus, but are for the most part still on the cusp of commercial success, include the use of industrial inkjet deposition in life sciences applications (such as proteomics, DNA sequencing, or even printed scaffolding for the growth of live tissues);[3] 3D rapid prototyping;[4] optical implementations such as lenses,[5] light pipes, and films; and electronic applications such as flexible displays, manufacture of color filters, conductive backplanes, LCD functional layers, spacer beads, black matrix, and printed electronics[6] including RFID, sensors, solar panels, fuel cells, batteries, and circuits.

Various technologies implement inkjet for varying reasons. Some examples:

Application	Benefit of Inkjet
Automotive coatings	Replaces spraying or dipping, thereby reducing waste and increasing coating uniformity.
Plastic part decoration	Non-contact accommodates curved surfaces. Improved print quality over pad or screen printing. Digital eliminates requirement for inventory of screens or pads, resulting in faster prototyping and a wider variety of designs. Process color capability reduces the number of ink colors that must be stocked.
Conductive patterns	Minimizes waste of costly materials; very suitable for short runs.
Rapid prototyping	Rapid formation of three-dimensional structures designed by using computer software.
Variable information	Allows fast changing of the printed information, unlike analogue printing methods which require formation of new hardware (e.g., screens in silk screen printing).
Ceramics	Minimizes setup time, eliminates requirement for inventory of screens.

Industrial Inkjet Explained

While all inkjet technologies can fundamentally be described as the digitally controlled ejection of drops of fluid from a print head onto a substrate, this is accomplished in a variety of ways. Industrial inkjet is broadly and most typically classified as either continuous (CIJ) or drop-on-demand (DOD), with variants within each classification.

As the name implies, continuous inkjet technology ejects drops continuously[2] (Fig. 1A). These drops are then either directed to the substrate or to a collector for recirculation and reuse. Drop-on-demand technology ejects drops only when required[7-9] (Fig. 1B).

Continuous Inkjet (CIJ) is considered amateur technology. It is primarily used for marking and coding of products and packages. In this technology, a pump directs fluid from a reservoir to small nozzles that eject a continuous stream of drops at high frequency (in the range of roughly 50 kHz to 175 kHz) by way of a vibrating

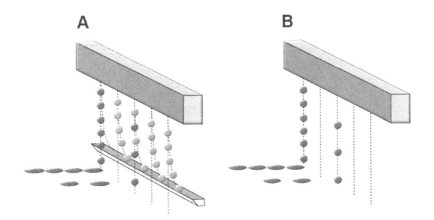

Fig. 1. Schematic representation of: (A) Continuous Inkjet (CIJ) and (B) Drop-on-Demand Inkjet (DOD).

piezo crystal. The drops are subjected to an electrostatic field to impart a charge, the charged drop then passes through a deflection field, which determines where the drop lands. Unprinted drops are collected and returned for reuse. The high drop frequency of CIJ directly translates to high speed printing capability as evidenced by such applications as the date coding of beverage cans. An additional benefit of CIJ is the high drop velocity (of the order of 25 m/s) that allows for relatively (compared to other inkjet technologies) large distances between the print head and the substrate, which is useful in industrial environments. Finally, historically, CIJ has enjoyed an advantage over other inkjet technologies in its ability to use inks based on volatile solvents, allowing for rapid drying and aiding in adhesion on many substrates. Disadvantages include relatively low print resolution, notoriously high maintenance, and a perception that CIJ is a dirty and environmentally unfriendly technology due to the use of volatile solvent-based fluids. Additionally, there are limitations associated with the requirement that the printed fluid has to be electrically chargeable.

Drop-on-Demand Inkjet (DOD) is a broad classification of inkjet technology where drops are ejected only when required. In general, the drops are formed by the creation of a pressure pulse.[7–9] The particular method used to generate this pressure pulse is what defines the primary subcategories within DOD. The primary subcategories are thermal,[10,11] piezo,[7–9] and electrostatic.[12] Sometimes, an additional category is discussed (MEMS), but MEMS drop-on-demand print heads are invariably still based on either piezo or thermal inkjet technology.

Thermal inkjet is the technology most used in consumer desktop printers and is making inroads in industry. In this technology, drops are formed by rapidly heating a resistive element in a small chamber containing the ink (Fig. 2). The temperature of the resistive element rises to 350–400°C, causing a thin film of ink above the heater to vaporize. This vaporization rapidly creates a bubble, causing a pressure pulse that forces a drop of ink through the nozzle. Ejection of the drop then leaves a void in the chamber that is subsequently filled by replacement fluid in preparation for creation of the next drop.

Advantages of thermal inkjet include the potential for very small drop sizes and high nozzle density, which leads to compact devices and lower print head and product costs. Disadvantages are primarily related to limitations of the fluids that can be used. Not only does the fluid have to be something that can be vaporized (implying most generally an aqueous solution), but it must withstand the effects of ultra high local temperatures. With a poorly designed fluid, these high temperatures can cause a hard coating to form on the resistive element, which then reduces its efficiency and, ultimately, the life of the print head. The high temperature can also cause problems if, for example, the functionality of the fluid is damaged due to the high temperature (as is the case with certain delicate fluids and polymers).

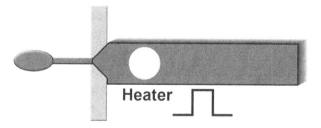

Fig. 2.   Schematic representation of a thermal print head.

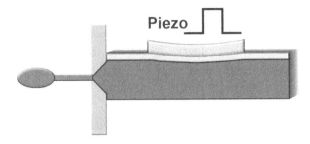

Fig. 3. Schematic representation of a Piezoelectric print head.

Piezoelectric inkjet is currently the technology of choice for most emerging industrial applications. In this technology, a piezo crystal (commonly lead zirconium titanate) undergoes distortion when an electric field is applied, and this distortion is used to mechanically create a pressure pulse that causes a drop to be ejected from the nozzle (Fig. 3). There are many variations of piezo inkjet architectures including tube, edge, face, moving wall, and piston.

Advantages of piezo inkjet technology include the highest level of ink development freedom of any inkjet technology, and long head life. Disadvantages include higher cost for print heads and associated hardware, limiting cost effective integration in low-end products.

There are very few commercial implementations of electrostatic inkjet, though these are increasing. Electrostatic inkjet (of which the most widely known is Tonejet by TTP) is characterized by drops being drawn from an orifice under the influence of an electrostatic field. This field, acting between an electrode and the orifice, attracts free charges within the ink (sometimes described as a liquid toner) to its surface in such a way that a drop is produced when the elec-trostatic pull exceeds the surface tension of the ink (Fig. 4). As this technique relies on the attraction of free charges, the ink is required to be conductive.

The advantages of an electrostatic inkjet are that it allows you to print a more concentrated fluid than the formulation that actually passes through the print head, and that the achievable resolution is not a function of the nozzle diameter so that potentially higher resolutions than piezo inkjet are possible. Additionally, very small drops can be formed while still using pigments, as the size of the drop

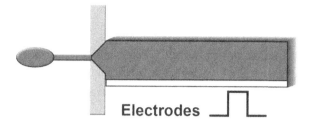

Electrodes

Fig. 4.   Schematic presentation of an electrostatic print head.

is controlled by the voltage on an ejection point and the properties of the particles, rather than by the size of the nozzle. As the printed material is significantly concentrated in the ejected drops, there is also the potential for high optical density images. Disadvantages include the limitation of only being able to use conductive fluids and the high cost of implementing the technology. As implementation increases, the cost is expected to go down.

There are literally thousands of companies participating in the design and/or delivery of industrial inkjet systems. Some are extremely vertically integrated (e.g., Hewlett-Packard) and are able to provide most or all parts of the complete solution, from ink to hardware to system integration to distribution, while others are focused on a particular aspect of the value chain (e.g., Xaar). The chart below gives a non-exhaustive indication of some of the major players in the various industrial print head technology variants. It should be noted that while MEMS print heads typically adopt either a piezo or thermal inkjet configuration, here they are shown separately due to the significant implications for the future of inkjet progress.

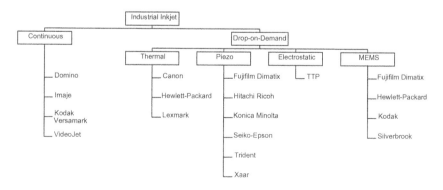

Inks

In achieving specific printing applications, the whole printing system should be evaluated, namely the print head, the fluid that is jetted from the print head (inkjet ink), and the substrate onto which the ejected droplets are placed. For applications, the requirements are established, defining the type of fluid chemistry, which directs the selection of the hardware and drives the implementation.

There are currently four main types of inkjet inks: phase-change,[13] solvent-based,[14] water-based,[15] and UV curable.[16] Other types exist, but are less prevalent, such as oil-based and liquid toner (for electrostatic inkjet technology). Hybrid versions of the four main types also exist (e.g., water-based inks containing some amount of solvent). The various inkjet ink types will be discussed briefly in this chapter, and will be followed by detailed description in separate chapters (solvent-based, water-based, and UV curable inks).

Phase-change inks, also known as hot melt, are distributed in solid form and, when introduced into a compatible system, are melted before being inkjet printed. Advantages of phase-change inks include that they are very fast drying (solidifying), environmentally friendly, and exhibit good opacity. It is also relatively easy to control the quality of the print because they do not tend to spread, due to their rapid solidification. Their primary disadvantages are the lack of durability and poor abrasion resistance. Phase-change inks are currently used in applications such as printing of barcodes on non-porous substrates.

Solvent-based inkjet inks have been around for many years and have traditionally been the formulation of choice for grand format and wide format applications due to exceptional print quality, image durability, and range of compatible substrates. They are also generally perceived as low cost. Benefits include the ability to adhere to a variety of substrates and fast drying time (which is often accelerated by heating). Solvent inks can be formulated with either pigments or dyes (or less commonly, both). Disadvantages include environmental concerns and a requirement for high maintenance, due to the potential of the fast drying fluid blocking the print head nozzles.

Water-based or aqueous inks are prevalent on the desktop and enjoy the advantage of being relatively inexpensive and environmentally friendly, but penetration in industrial applications has been slow for a variety of reasons. Water-based inks tend to require porous or specially treated substrates or even lamination to impart durability and the ink tends not to adhere to non-porous substrates. Additionally, many piezoelectric industrial print heads are incompatible with water-based ink formulations, although this is changing in some part due to market demand for systems that can jet water-based biological or food contact fluids.

UV curing chemistry for inks and coatings has been used in printing markets for many years, and thanks to recent investment in the R&D of inkjet print head and fluid formulation, inkjet is now an established and robust deposition tool for UV curable fluids. This is not surprising considering the benefits brought about by the partnering of UV and inkjet technology. UV inks are designed to remain as a stable liquid until irradiated with a particular wavelength and intensity of light.

UV inks are now reliably and successfully employed for a variety of inkjet applications across many different sectors. The benefits afforded by UV, coupled with the flexibility of digital printing, have proved a compelling proposition for many industrial applications and have seen the breadth of UV ink implementations spread from the more traditional wide format/flatbed sectors into the niche application areas of product coatings, primary package decoration, and labelling. Current limitations are in edible and food contact applications. Disadvantages include cost and facility requirements (space, extraction, power) for the UV curing hardware.

As stated earlier, inkjet printing is a system, which should take into consideration the hardware, the ink properties, and the interaction with the substrate. Once the requirements are defined and the ink chemistry and inkjet printing technology have been chosen, there are additional considerations including image/information processing, speed, print quality, cost trade offs, fixed vs. scanning heads, and maintenance systems. In the case of graphics printing, the optical properties of the colorants play an essential role in the final perception of the image. The image is actually a combination of process colors[17] — cyan, magenta, yellow, and black — and, therefore, the placement of each ink drop and the order of placement, as well as bleeding issues, play significant roles in the print quality. In recent years new inkjet systems have been developed to include, beyond the CMYK set, light magenta, light cyan, and white to widen the color gamut.

For emerging materials deposition applications such as printed electronics, inkjet system requirements are diverse and can include ultra-high precision substrate handling, drop visualization, and fiducial recognition for printing of multiple layers, not to mention the requirement for inkjet fluids that may incorporate "difficult" ingredients such as nano or large particles that must remain in suspension, aggressive acids or alkalis, fragile biological materials, magnetic materials, and in some cases, even radioactive substances.

Technology Trends

Innovation in industrial inkjet is fast and furious. As an indication of the technological activities in this field, there were 3553 US and European patents and patent applications in 2006 alone (translating to roughly 300 per month), making inkjet one of the most actively patented technologies in the world. Hewlett-Packard, Canon, Seiko Epson, and Silverbrook lead the patenting pack, but their efforts cover less than half of all the current activity.

In the past, the primary focus of new inkjet technology development was in the increase of print resolution. By the use of smaller and more accurately placed drops, clever image processing/manipulation and greyscale techniques, inkjet has reached the limit of what the human eye can differentiate (evidenced by today's low cost, ultra-high image quality consumer printers).

Today, emphasis is placed on throughput improvements by way of increases in raw jetting speed as well as inline, single pass implementations; reliability improvements through the development of self-recovering print heads, integration ease and scalability resulting in elegant and lower cost industrial implementations, development and extension of pre- and post-processing techniques (such as e-beam curing and UV LED pinning) to extend the capabilities of inkjet, and last, but certainly not least, the enabling of new applications through smaller drop sizes, increasingly accurate drop placement (fueled by the adoption of MEMS technology) and new fluid developments.

Most existing inkjet implementations are multipass, where single or multiple print heads move backwards and forwards across a substrate, building up an image. This can offer high print quality since multiple passes can be engineered to mask the effects of blocked nozzles, but this comes at the cost of speed. In single pass configurations, one or more print heads cover the entire width to be printed. This has great potential for higher throughput, but has historically presented reliability, manufacturing, and cost challenges. These challenges are being addressed and a number of companies are introducing, or announcing, single pass solutions. In particular, the Xaar 1001 print head has been specifically designed for single pass printing; Fujifilm Dimatix announced its SAMBA technology — a single pass prototype head — in May 2008; and Kodak's Stream technology, targeted for shipment in 2010, is a single pass solution that claims to have the print quality and speed of offset lithography. Single pass solutions are ideal for web-based applications such as label printing or any application requiring high throughput. Hewlett-Packard has announced a web press targeted at newspaper and digital book printing that has a speed of 122 m/min at 600 dpi (shipping toward the end of 2009) and Kyocera recently announced availability of what is reportedly the world's fastest high resolution piezo print head with a top speed of 150 m/min at 600 dpi.

Another vector of development in industrial inkjet printing is reliability. It is not unusual for piezo drop-on-demand industrial print head nozzles to achieve successful lifetimes in excess of 10^{13} drops, but this masks the real world requirement of 24/7 operation since print heads can sometimes require a significant amount of ongoing maintenance including purging and cleaning. For example, Xaar is addressing this with the implementation of "self recovery" techniques in the 1001 print head, in which a continuous ink flow through the channels at $10\times$ the flow rate through the nozzles provides a quick recovery from air ingestion and increased operational time between maintenance of hours rather than minutes.

Other attempts to improve system reliability include implementation of vision systems to detect misfiring nozzles, increasingly being incorporated in manufacturing tools for functional printing (such as electronics or bio) because these applications typically require perfect deposition to ensure functionality.

Partly due to the large amount of attention being put on replacing existing manufacturing technologies with inkjet deposition, there is increasing focus on providing scalable print head technologies that can be reliably and economically integrated. While pursuing their quest for the ideal print head, Silverbrook has developed a technology that is highly scalable from the standpoint of width.

Even mature applications, such as grand format printers, continue to pursue ever increasing print widths in a scalable fashion though not necessarily economically. As an example, Inca Digital is offering the Inca Onset which includes an array of 576 Dimatix print heads (translating to 73 728 nozzles) situated in plug-in print bars with an innovative alignment system.

With the push for smaller feature sizes to enable the benefits of inkjet printing in functional applications, such as printed electronics,[18] a great deal of effort is going into developing technologies that can produce ever smaller drop sizes. As drops get smaller, the energy needed to eject them from the print head must increase so that the effects of fluid surface tension can be overcome. Additionally, as drops become smaller, their surface area to mass ratio changes and, as a result, they tend to decelerate more quickly, which reduces the allowable throw distance. These challenges impact both print head design and fluid formulation. The smallest current drop size for production is technology from FujiFilm Dimatix at 1 picolitre. Today, a print head with the smallest drop size coupled with an optimized fluid formulation, coupled with the perfect ink/substrate combination, coupled with ultra-precise substrate handling, is likely to lead to consistent spot sizes of roughly 30 microns, with sizes as low as 10 microns in laboratory settings.

To further reduce feature sizes, there are a number of non-inkjet techniques presently under investigation, such as self-aligned printing and surface energy patterning. Some predictions suggest that feature size will get down to as low as 10 microns in high volume manufacturing in as few as 5 years.

Many of these technological advances are enabled by the use of IC manufacturing techniques to produce ever finer print head features. DRIE (Deep Reactive Ion Etching) also allows for near vertical walls. Other benefits of MEMS fabrication methods include submicron accuracy, robust materials, and the ability for high volume, low cost manufacturing. MEMS techniques are ideal for the creation of nozzles, manifolds, and channel structures in inkjet print heads. Hewlett-Packard is a pioneer in using MEMS for the manufacture of print heads, and Silverbrook has only ever offered MEMS print heads. Fujifilm Dimatix offers an M-class of print heads that take advantage of MEMS technology. The silicon nozzle plate of these heads is much more resistant to scratching than other piezo print heads, and it also offers ultra-precise directionality of ink drops.

Inkjet developments are not limited to inkjet technology. Great success has been made with the combination of UV curing and inkjet, and this is being expanded to related technologies such as e-beam curing and low cost LED (Light Emitting Diode) UV for pinning. In one example, UV curing has limited use in food or food contact related applications due to the requirement of photoinitiators in the formulation. These photoinitiators can be toxic. E-beam curing, which does not require photoinitiators in the ink, is being considered as an alternative, having the advantages of UV curable inkjet (adhesion, abrasion resistance, print reliability, high speed) without the disadvantages. However, the price of the e-beam equipment is currently a limiting factor. In the case of LED UV, these devices are low power, low cost, and do not generate much heat. They can be used to "freeze" rather than fully cure printed drops immediately upon impact with the substrate. This allows precise control of substrate wetting and print quality and can significantly aid throughput and/or print quality when multiple fluid types are required.

Of all these trends, arguably the most important for adoption of inkjet technology is increasing the range of jettable fluids. As an example, in textile printing applications, the use of sublimation dyes that volatilize at high temperature to migrate and bond strongly to the textile fabric to produce a water washable robust image are less than desirable due to the ancillary heating and washing processes required. Utilization of pigmented textile inks can remove some of these process requirements as well as be more suitable for a wider range of textiles (natural and man made). Other examples, such as printing of ceramic inks, direct printing of conductive patterns by using metallic nanoparticles, and printing of 3D plastic structures, will be discussed in separate chapters. Obviously, a main area of activity in formulation of inkjet fluids is the conversion of existing non-inkjet fluids to inkjet, by adjusting the physicochemical properties of the liquids to the overall inkjet system requirements.

Challenges

Inkjet must be understood as a complete system. Many disciplines (materials science, chemistry, device physics, system integration, production engineering, software, mechanical engineering, electronics) must be brought together. Customers still face application challenges and available production tools, still in their infancy, often do not meet all requirements. It is not uncommon for systems to exist that work in the laboratory but are not yet ready for a 24/7 industrial environment.

The main challenges in improving the performance and utilization of inkjet printing are:

Materials: Increasing, but still slow, is the development of jettable materials. There is no such thing as a universal ink. In each case, many issues have to be considered, among them: application performance (functional), print quality (bleed, surface wetting), compatibility, drying/curing time (relating to speed), adhesion (sometimes interlayer interactions), image robustness (water fastness, gas fastness, light fastness, abrasion resistance) jetting characteristics (viscosity, dynamic surface tension, particle size, compatibility), reliability (volatility, wetting of capillary channels, priming, purging, shelf life), ease of manufacturing (milling, cost and availability of materials), regulatory (FDA, etc.), post- or pre-processing requirements (UV, e-beam, heating, inert atmospheres).

An abundant approach to achieving jettable materials is based on adjusting the composition of an existing ink to match the printing system requirements. Usually this approach is not trivial, since non-inkjet formulations usually have very different properties to inkjet inks. For example, converting a silk screen printing ink would require, among other changes, a significant decrease in the viscosity, and a significant decrease in particle size in pigment-containing inks. Decreasing the viscosity, in the case of large pigment particles, would lead to sedimentation and aggregation of the particles. To prevent this would require a submicron pigment size (also important for not clogging the print head). Such pigments (metallic, ceramic, etc.) are not always available commercially, and in that case they should be manufactured specifically for the new inkjet ink.

Feature size: Another challenge is associated with feature size reduction, especially for sophisticated printing of functional materials,[19] such as in printed electronics. This can be achieved by combined effects of the whole printing system, such as surface treatment of the substrates (see separate chapter) and achieving unique rheological behavior of the ink.

Resolution and productivity: Higher resolution and substrate handling at higher speed is a very demanding task. While approaching the fundamental limits of increased jetting frequency, the productivity needs to be improved in other creative ways. To date, this has been accomplished through increasing the number of nozzles, although this is directly related to increased cost.

Drop placement accuracy: Exact drop landing position is uncertain, due to various parameters such as jet-to-jet variations, single jet-over-time, sensitivity to nozzle straightness, nozzle and surface wetting, nozzle plate contamination, ink formulation and condition, and drop velocity. This issue is worse for longer flight paths or "throw distance".

In summary, inkjet success is based on treating the "inkjet detail" with respect. Although it is elegant in concept, it is very difficult to implement in practice, especially in very demanding applications such as very high throughput systems, and printing of functional and unique materials.

REFERENCES

1. Elmqvist R. (1951) US Patent No. 2,566,443.
2. Sweet R. (1971) US Patent No. 3,596,275.
3. Sachlos E, Wahl DA, Triffitt JT, Czernuszka JT. (2008) The impact of critical point drying with liquid carbon dioxide on collagen-hydroxyapatite composite scaffolds. *Acta Biomater* **4**(5): 1322–1331.
4. Napadensky E. (2003) U.S. Patent No. 6,569,373.
5. Momma T. (2006) European Patent No. EP 1683645 A1.
6. Sirringhaus H, Tatsuya S. (2003) Inkjet printing of functional materials. *Materials Research Society Bulletin* **Nov**: 802–806.
7. Zoltan S. (1972) US Patent No. 3,683,212.
8. Stemme N. (1973) US Patent No. 3,747,120.
9. Kyser E, Sears S. (1976) US Patent No. 3,946,398.
10. Kobayashi H, Koumura N, Ohno S. (1981) Liquid recording medium. US Patent No. 4,243,994.
11. Buck RT, Cloutier FL, Erni RE, Low RN, Terry FD. (1985) US Patent No. 4,500,895.
12. Silverbrook K. (1998) US Patent No. 5,781,202.
13. Berry JM, Corpron GP. (1972) US Patent No. 3,653,932.
14. Bhatia YR, Stallworth H. (1986) US Patent No. 4,567,213.
15. Chandrasekaran CK, Ahmed A, Henzler TE. (2003) European Patent WO 2003076532 A1.
16. Noguchi H, Shimomura M. (2001) Aqueous UV-curable ink for inkjet printing. *RadTech Report* **15**(2): 22–25.
17. Pond SF, Wnek WJ, Doll PF, Andreottola MA. (2000) Ink design. In Pond SF (ed), *Inkjet Technology and Product Development Strategies*, pp. 153–210. Torrey Pines Research, Carlsbad.
18. Isobe M, Takiguchi H, Kiguchi H, Shibatani M. (2008) Japan Patent No. 2008233572 A.

CHAPTER 2

Ink Requirements and Formulations Guidelines

Shlomo Magdassi
The Hebrew University of Jerusalem, Israel

Due to the complex nature and very challenging requirements of inkjet inks, preparation of such inks is often very complicated. In addition to the conventional requirements, such as long shelf life and proper color properties, the ink must have physicochemical properties which are specific to the various printing devices. For example, each print head has a specific window of surface tension and viscosity range which enables proper jetting. Piezoelectric print heads usually function within a viscosity range of 8–15 cP, while thermal print heads perform at much lower viscosity, usually below 3 cP. The end use of the printing system also dictates the physico-chemical properties of the inks; for example, an ink developed for home and office use should print many pages without any operator technical maintenance, while for industrial printers, such mainte-nance is acceptable if it brings added value such as reducing energy requirements during drying of the ink on various substrates. This means that selection of the volatile components of the ink, in the case of water- or solvent-based ink, can be affected tremendously by the end use and the printing environment.

Therefore, while formulating new inkjet inks, the formulator must take into consideration the effect of each component on the overall performance of the ink, from storage in the ink cartridge, through jetting, to its behavior on the substrate and its effect on health and the environment. Such considerations are valid for all types of inkjet inks (solvent, water, reactive, etc.) and also for inks which bring a sophisticated property, beyond graphic performance, such as conductive inks. The latter is a good example of a functional ink, which should meet the usual inkjet ink requirements, but in addition should provide good electrical conductivity. Obtaining such a functional property often presents conflicting directions for the scientists who prepare the inks: in order to obtain a stable ink, which is composed of metallic nanoparticles, the best way to stabilize the metallic dispersion, as will be discussed, is by using a charged polymeric stabilizer which provides an electrosteric barrier against coagulation of the particles. This is essential since the density of metallic particles, such as silver, is much greater than that of water and typical liquid vehicles of inkjet ink, and therefore flocculation followed by sedimentation is more problematic compared to conventional organic pigments (in addition to the usual orifice clogging issues). However, in order to obtain good electrical conductivity after printing, the metallic particles should form a continuous path by making contact between the particles. This is difficult to achieve due to the presence of the polymeric stabilizer between the nanoparticles.

Another example of conflicting requirements is in UV inks: In order to obtain high throughput in an industrial printing system, the ink must be capable of undergoing rapid curing after UV radiation. Although the fast curing will provide good printing resolution and high throughput, it will not favor the spreading of the ink droplets over large areas, so the ink coverage will be low. Overcoming low coverage would require placing more ink droplets on the substrate, and obviously means higher ink consumption and cost.

In this chapter the principle issues which govern the overall performance of the ink will be discussed from the chemical point of view, and will be divided into three main groups of functional requirements based on the stage in the life cycle of the ink, from manufacturing and storage to its fixation on the substrate.

It should be emphasized that formulation of a new inkjet ink would require the integration of all three, in addition to issues such as environmental and health regulations.

INK PREPARATION AND COMPOSITION

The main function of ink is to place functional molecules on a substrate, after being jetted from a print head. The functional molecules can be colorants (pigments or dyes) as known in the graphic arts, a conductive polymer for printing "plastic electronics", a UV-curable monomer for printing a three-dimensional structure, a living cell for printing an organ, a polymer for a light-emitting diode,[1] etc. The common feature of all inks is that when they are jetted from the print head, the whole ink should be liquid, having, for most industrial print heads, a viscosity below 25 cP. Therefore, the ink can be regarded, in principle, as composed of a vehicle and a functional molecule. This vehicle is actually the liquid carrier of the functional molecules. This definition also holds for hot melt ink, in which the vehicle is solid before and after printing, but is liquid during printing.

As discussed in the following chapters, the vehicle is composed of the liquids (e.g., water, organic solvents, cross-linkable monomers), additives which bring a specific function (e.g., surfactant, preservative, photoinitiator) and usually also a polymer which enables the binding of the functional molecules (e.g., colorants, conductive particles) to the substrate after printing.

Typically, as shown in Fig. 1, the ink preparation is based on formation of a varnish, which is the liquid containing most of the ink

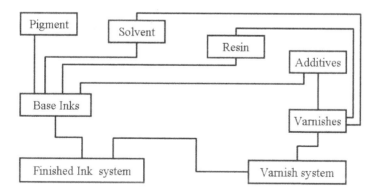

Fig. 1.   Inkjet ink preparation scheme.

components (polymeric binder and additives), mixed with the functional molecules, which are either insoluble (pigments) or soluble (dyes) in the liquid phase. It should be noted that the difference between a dye and a pigment-like functional molecule is related to its solubility in the liquid vehicle; a water-soluble dye can become a pigment if the vehicle is converted into an organic solvent.

The selection of the various components of the vehicle is tailored according to the printing technology and the final function of the printed pattern. It is common to divide the inks according to the nature of the liquid vehicle: water-based inks (main component of the liquid is water), solvent-based inks (liquid is composed of several organic solvents), reactive inks (main component of the liquid can undergo polymerization reaction after triggering, such as UV inks which polymerize by UV radiation).[2] The ink components and the so-called additives which are utilized in specific ink types will be discussed in other chapters of this book.

INK DURING STORAGE

Once the ink is prepared, it should meet specific physicochemical criteria, which depends on its intended use.[3] All the properties of the ink should remain constant over a prolonged period of time ("shelf life"), which is typically two years storage at room temperature, but currently there are exceptional shorter times, e.g., for UV inks. The main ink parameters that should be considered while preparing the ink and are mainly related to bulk properties of the ink are discussed below.

Ink Stability

A stable ink is an ink in which all its properties remain constant over time.[4] In most cases, for inks which do not contain undissolved materials, instability is caused by interactions between the ink components, such as polymerization in UV ink, precipitation and phase separation due to changes in solubility (encountered, for example, when samples are stored or shipped at low temperatures), and even interaction with the walls of the ink containers. For example, adsorption of a wetting agent from the ink onto the polymeric walls of the container may result in an increase in surface tension, or partial polymerization of monomers during storage could lead to an increase in ink viscosity.

For inks which contain pigments, the most common problem is aggregation of the pigment particles due to the inherent instability of most dispersion systems. Since most modern inkjet inks for graphic applications contain dispersed pigments, the stabilization mechanisms of dispersions will be briefly discussed below.

When particles approach each other, interaction due to van der Waals forces takes place, causing the particles to aggregate and eventually reach the minimum potential energy, as shown in Fig. 2. For example, if the pigment is hydrophobic, such as carbon black, it will tend to form large aggregates in water. In order to prevent the aggregation, a mechanism to overcome the attraction is required. Electrical repulsion, which can be obtained if the surface of the pigment particle bears electrical charges, is such a mechanism. For example, if an anionic surfactant, such as SDS, is adsorbed on the surface of hydrophobic pigments, it will impart negative charges

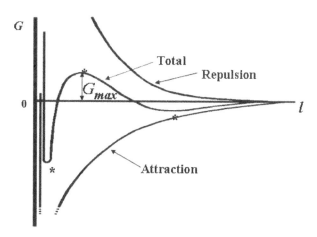

Fig. 2.   Total interaction energy curves of a colloidal system.

to the pigment surface. In this case, while particles approach each other, electrical repulsion will take place as the distance between the particles decreases (Fig. 2).

As described by the DLVO theory,[5] if the repulsion overcomes the attraction, an energy barrier will exist and prevent aggregation of the particles. As clearly seen in Fig. 2, the dispersion will be thermodynamically unstable, but if the energy barrier is sufficiently high, the system will be kinetically stable. Stabilization depends on various parameters such as the zeta potential of the particles, concentration and valency of the ions present in the solution, and the dielectric constant of the solution in which the particles are dispersed. The electrostatic stabilization mechanism is effective in systems having a high dielectric constant, and therefore is mainly important for water-based inks. Additional stabilization can be achieved by a steric mechanism, in which a polymer is adsorbed onto the surface of the pigment (through groups in the polymer which have affinity to the pigment surface), and provides steric repulsion. For example, with carbon black pigment, stabilization can be achieved using a polymer which has hydrophobic groups which can bind to its surface, and also has sufficiently long hydrophilic segments that are soluble in water. This stabilization mechanism is very effective for both aqueous and non-aqueous ink systems, and there is a large variety of commercially available polymeric dispersants, such as Efka, Tego Dispers, Solsperse, Disperbyk, Sokalan®.[6,7]

It should be emphasized that in order to achieve a stable dispersion of pigment-containing ink, one should evaluate various dispersing agents and find their optimal concentration. The selection should provide a dispersant which has anchoring groups enabling its adsorption on the pigment surface, while enabling fragments of the dispersant to be extended into the solution. The dispersant concentration is of great importance: there is an optimal concentration of dispersant, which can be determined by size and viscosity measurements. As shown in Fig. 3, the viscosity of the dispersion usually decreases with the increase in dispersant concentration, up to a certain concentration, followed by increasing viscosity above

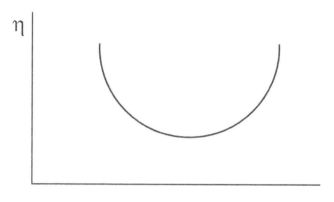

Dispersant concentration

Fig. 3. Typical dependence of ink viscosity on dispersant concentration.

that concentration. The latter increase can be a result of free polymeric dispersant dissolved in the liquid medium.

Stability is evaluated by measuring the various parameters of the ink for a prolonged period of time. However, for practical reasons these tests are conducted after storage at accelerated conditions, such as high temperature, low temperature, and freeze-thaw cycles. The accelerated storage conditions vary from laboratory to laboratory, since the correlation to storage in "real" conditions is not very precise and depends on various parameters.

Viscosity

The viscosity of the ink is of great importance for its performance during jetting and spreading on the substrate, and is affected by many parameters — among them, the presence and concentration of polymeric additives, surfactants, solvent composition and flocculation. Most current inkjet inks are Newtonian, i.e., they have a constant viscosity over a wide range of shear rates. The viscosity of inkjet inks is very low, usually below 20 cP, depending on the print head (below 3 cP for thermal print heads). Since the viscosity may increase due to flocculation of particles, viscosity measurements during storage would provide an early warning regarding aggregation of the pigment particles. Another common problem is the increase in viscosity during storage encountered in UV inks, which may undergo polymerization reactions during storage. In such cases, formation of oligomers is sufficient to cause a viscosity increase that would significantly affect the overall performance of the ink during printing. It should be emphasized that the window of operation, in terms of allowable deviations from the starting value in viscosity, is dependent on the system requirements, mainly print head performance.

Since the ink has a low viscosity, even at low shear rates, sedimentation of pigment would occur if the pigment particles are large and/or have a high density (such as metals or ceramics). Therefore, great effort is made to obtain particles as small as possible. Another stability-related possibility regarding ink viscosity is the use of ink which has a high viscosity at room temperature and low viscosity during jetting (at elevated temperatures). The extreme example of such ink is the hot melt ink, which is a solid at room temperature and a low viscosity liquid above its melting point. By using this concept, sedimentation of pigment particles during storage is prevented. Taking advantage of the very different shear rates encountered during storage and during jetting opens the possibility of a non-Newtonian ink,[8] namely, an ink with pseudoplastic behavior: during storage, the ink will have a high viscosity (low shear rates), and during jetting, low viscosity (very high shear rate, in the range of $1\,000\,0000\ S^{-1}$). Obviously, such behavior would affect other ink properties such as ink flow through the printing system, jetting and drop breakup.

Surface Tension

The surface tension of the ink is a primary factor determining droplet formation and spreading on the substrate upon contact.[9] The surface tension can be controlled by using surfactants and by selecting proper solvent compositions. For example, adding propanol to water will cause a large decrease in surface tension, from 72.8 dyne/cm to below 30 dyne/cm, depending on the propanol concentration.

Significant decreases in surface tension with the addition of a co-solvent are usually obtained at relatively high co-solvent concentrations. A surfactant is usually used at very low concentrations, sometimes much below 1% w/w, very often even below 0.1% w/w. This means that even a slight change in the surfactant concentration may cause a significant change in the ink performance. Such a decrease can result, for example, by adsorption of the surfactant on the ink container walls or on pigment particles during prolonged storage.

If the surface tension results from the composition of the liquid medium, it will not change over time, and the surface tension value will be that of equilibrium conditions, which can be readily measured by conventional methods, such as the du Nouy ring or Wilhelmy plate method.[10] However, if the surface tension is controlled by surfactant, the dynamic surface tension should also be considered. This parameter is important in cases in which a new surface is formed (such as during drop formation or spreading of a drop on a substrate) and is not covered yet with the surfactant molecule; thus initially it has a high surface tension. After diffusion of the surfactant to the interface, the surface tension will decrease, until it reaches its equilibrium value. The dynamic surface tension can be measured by a variety of techniques, such as the bubble pressure method.[10] It should be emphasized that the resulting surface tension (static and dynamic) depends on all the components present in the ink and could be affected by interactions between the components, such as binding of surfactant to dissolved polymer[11] and even due to migration of a plasticizer from the ink container.

pH and Electrolytes

The pH is important in water-based inks and may significantly affect the solubility of the various components and the stability of the dispersed pigments. The solubility effect is often observed when the ink contains a polymeric binder such as acrylic resin, which is insoluble at low pH. As discussed above, colloidal stability is affected by the zeta potential of the particles: the higher the value, the higher the stability. In many cases stabilization is achieved by adsorbed polymers, which are effective while they are charged; the charge is usually dependent on the pH of the system. Therefore, some ink formulations contain buffers, which make the ink less vulnerable with regard to slight variations in ink components and water quality.

The presence of electrolytes can cause severe stability problems during prolonged storage, due to compression of the electrical double layers of the particles (which may cause flocculation); therefore the concentration of electrolytes must be very low. This is especially important for multivalent electrolytes, such as calcium. Consequently, it is essential to control the water quality, and some formulations even contain sequestering agents such as EDTA.

Dielectric Properties and Conductivity

These properties are essential for continuous inkjet inks, in which the droplets are deflected due to an electrical field. The charging ability is obtained by adding charge control agents — electrolytes and ionic surfactants — which are soluble in the ink medium. The conductivity should be very precisely controlled; therefore any slight variation in conductivity during storage should be prevented. Here, too, variation in conductivity may occur due to salt precipitation, interactions with other components and the wall of the container, etc. The ink conductivity is also important for printing systems in which ink recirculation sensors are triggered by conductivity signals obtained by contact of the sensor with the ink.

Dye/Pigment Content

The main function of the ink is to bring a functional molecule, usually a colorant, to a substrate. If the colorant is a dye molecule (or combination of various molecules) it should be present at a concentration much below its solubility limit, otherwise slight variations during storage (e.g., temperature, pH) could cause precipitation. In such inks it is essential to determine the dye solubility in presence of all the components, especially at low temperatures. The optical properties of dyes are often affected by slight variations in pH and presence of electrolytes (in water-based inks), medium polarity, and presence of surfactants (possible solubilization). Therefore, the optical properties during prolonged storage should be tested, either in the liquid form or by draw-downs.

In general, dye-containing inks are more stable than inks containing pigments, since the ink is thermodynamically stable (all components are dissolved in one solution), while in pigment inks the system is only kinetically stable. Conventional inks usually contain pigment at concentrations below 10% w/w in order to achieve the required optical density. Non-graphic inkjet ink, such as conductive or ceramic inks, may contain pigments at concentrations greater than 50%. In general, the higher the pigment loading of the ink, the greater the chances for contacts between particles and aggregation. However, it should be noted that in such pigments, due to their high density, the volume fraction (which affects the stability) is much lower. The main problem related to the bulk behavior of pigment-containing inks is the possible aggregation (and eventually sedimentation, since most inks have low viscosity), which would cause a decrease in optical density after printing, and may clog the print head nozzles during jetting. Therefore, great efforts should be made to prevent aggregation of the pigment particles during prolonged storage. If the aggregation is irreversible, the ink usually cannot be used (unless reprocessed), but if the aggregation is reversible, sometimes shaking and circulation within the printer will be sufficient. Evaluation of the aggregation issue is performed by particle size measurements (dynamic light scattering, etc.), by filtration tests, and by evaluating the optical density losses after storage.

Foaming and Defoamers

A severe problem in ink performance is the presence of bubbles in the ink.[9] Foaming is often observed in inks which contain surfactants and polymers. A common solution to this problem is addition of a defoamer, which is a molecule that causes breakdown of foam which is already present. The defoamers act by: a) reducing surface tension in a local area to very low values, causing these local areas to be thinned rapidly (example: amyl alcohol); b) promoting drainage of liquid from the lamellae (example: tributyl-phosphate which reduces surface viscosity).

Additives performing through the first mechanism should have limited solubility in the ink, and usually contain an immiscible moiety, e.g., a silicon derivative. Obviously, such molecules will tend to separate out of solution during prolonged storage, but since they are present at very low concentrations, it will be difficult to detect visually. The effect of phase separation of insoluble material can be very significant during jetting (changing the wetability of the print head) and after printing (forming surface defects caused by low surface tension spots). Therefore, one should avoid, if possible, the use of defoamers. If it is essential due to severe foaming of the ink, one should seek defoamers which do not separate out during prolonged storage, and use concentrations as low as possible.

INK-PRINTHEAD PERFORMANCE

Once the ink meets the physicochemical properties which are required for a specific print head,[12] it should be tested for jetting and print head performance. Here the focus is on the general technical requirements which are crucial for obtaining an ink with good jetting performance, from the formulation view point.

Drop Latency and Recoverability

Conventional water-based and non-aqueous inkjet inks are mixtures of several components, including volatile solvents, dissolved materials, and dispersed solids (for pigment inks). When the ink reaches the nozzles prior to jetting, the volatile components may evaporate from the nozzle.[3] Therefore, the liquid in the vicinity of the nozzle can have a composition which differs from that of the bulk ink which is further back in the print head supply channels. This disparity causes differences in the physicochemical properties of the ink (e.g., an increase in viscosity or decrease in surface tension) that may result in shifting away from the required properties for proper jetting; thus a drop cannot be jetted after prolonged idle time ("first drop problem"). The time that inkjet ink can successfully wait in an individual orifice, without jetting, is termed **latency**.[13] This period varies from a few seconds to a few minutes in commercial Drop-on-Demand water- and solvent-based inks, and reaches many days for inks which do not contain volatile solvents, such as 100% UV inks.

If significant evaporation of solvents occurs at the nozzle, the solubility of some of the components may be exceeded, resulting in precipitation of those components. This means that there are crusty deposits which can cause a blockage of the nozzle. A similar phenomenon can be found if the concentration of pigment particles at the orifice increases, leading to severe aggregation due to the increase in volume fraction of the pigment in the ink. A similar situation may result in inks in which the solubility of the components or stability of pigment depends on the pH: if the ink is made basic by addition of a volatile amine, its evaporation may cause a decrease in pH, followed by crust formation. Less problematic, but still significant, is the possibility of increased viscosity due to evaporation, which is often encountered in inks which contain polymeric binders. This would cause a "viscous plug" that may prevent jetting or result in small and/or slow drops. The crust can either completely block the jetting or cause a misdirection of the jetted drops.

To obtain good latency, the ink formulation tools are those which decrease the evaporation rate and form an ink which is less vulnerable to evaporation of some of the solvent. These tools are:

a. For solvent-based inks, use less volatile solvents, solvents with higher boiling points and lower evaporation rates (the evaporation rate is usually given relative to butyl acetate).
b. For water-based inks, add co-solvents capable of delaying the loss of water by hydration. Typical co-solvents are glycols such as diethylene glycol, polyethylene glycols, and propylene glycol methyl ethers (Dowanols®). Such additives are often termed "humectants".

c. For water- and solvent-based inks, maximize the solubility of solids in the liquid by selecting solvent composition and co-solvents (e.g., N-methyl pyrrolidone for water-based inks) which enable dissolution even after a fraction of the liquid medium has evaporated.

d. For pH sensitive inks (required to keep high solubility or high zeta potential), use non-volatile pH control agents. For example, if the ink requires a high pH for dissolution of an acrylic polymer, use an amine which has a very high boiling point.

e. For pigment-containing inks, select polymeric stabilizers which would keep the viscosity low even at high pigment load.

f. For inks containing polymeric binders and rheology control agents, select those in which their solution viscosity is less susceptible to changes in pH or polymer concentration (e.g., low molecular weight, proper acid number).

Recoverability

Since any ink must eventually be fixed and dried on the substrate (except hot melt and UV inks), it should have some volatile components, so from time to time the ink will not be jetted after idle periods. Therefore, the inkjet ink and ink system should be capable of restoring the ink in the nozzle region to the initial composition. This is achieved by ejecting ink into a waste collector ("spittoon") in the maintenance station. The ink formulator should make sure that the ink has the capability to redissolve the ink crust, and is able to redisperse aggregated particles. This ability can be preliminarily tested in the laboratory by drying a small sample of ink, and testing how fast the dried ink returns to its original properties upon addition of fresh ink. The formulation tools are essentially similar to those related to the latency issues, focusing on those ink components affecting dissolution and aggregation.

In practice this test is also performed by measuring the quantity of ink that has to be ejected until proper jetting is enabled, and is termed **recoverability**.

Orifice Plate State

Another issue which can cause jetting problems is the wetting state of the nozzle faceplate. For example, if the drop formation process and jetting are not optimal, some of the ink may accumulate on the orifice faceplate. If the surface tension of the ink is sufficiently low, the ink will be spread as a thin layer on the orifice plate, and if evaporation occurs, a solidified layer on the orifice plate may be formed and prevent jetting, depending on the print head jetting mechanism and properties of the wetting film. The wetting of the orifice plate depends on the surface energy of the orifice plate material, and the surface tension of the ink. As an empirical rule, good wetting occurs while the surface tension of the ink is lower than the surface energy of the faceplate material. Therefore, the formulation tools control the surface tension of the ink by proper selection of the type and concentration of wetting agents such as fluorosurfactants,[14] and the composition of the liquid components, which affect both the surface tension and evaporation rate of the liquid medium.

Ink Supply and Clogging

In order to be jetted properly from the print head, the ink must go from the ink container (or cartridge) to the print head, passing through tubes and various filters. The two main parameters which affect the flow and the filtration issues are the rheology of the ink and the particle size in pigment-containing inks. Most of the inks are Newtonian and have sufficiently low viscosity to enable the flow of inks through millimeter-size diameter tubes that are part of the ink supply system. However, there are inks which are designed to be jetted at high temperatures, thus at low viscosity, while the viscosity is higher at room temperature. In such cases, it is crucial to also adjust the room temperature viscosity, so that sufficient ink arrives at the print head; otherwise, a starvation phenomenon would occur, especially at high jetting frequencies, in which the ink consumption is high.

In rare ink formulations, the ink may have a non-Newtonian rheology, i.e., high viscosity at low shear rate and low viscosity at high shear rate (shear thinning inks). In such a case, once the ink reaches the orifice it may be jetting well, since the shear rate in the nozzles is very high, but it may not flow properly in the ink supply tubing and the narrow channels of the print head, in which the shear rate is low.

The ink supply system and most print heads have filters which are aimed at preventing arrival of large particles to the nozzles. Aggregation of pigment particles usually causes an increase in viscosity, which can interfere with the ink flow through the ink supply system.[6] The aggregates can block the filters and thus may decrease the flow rate over time, eventually causing starvation of ink in the print head.

From an ink formulation point of view, there are two main parameters which are crucial for obtaining a good flow of the ink:

a. Achieving a very stable ink with particle size less than 200–300 nm (rule of thumb: the orifice diameter should be about 100 times the particle diameter. If the orifice diameter is about 40 μm, the particle size should be less than 400 nm. There are several reports on methods to prepare inkjet pigments having a D50 below 20 nm; such inks should have good performance regarding clogging issues (if they are stable).[15,16] Large particles or large aggregates formed during ink storage may escape the filters on the ink supply system, but eventually reach the nozzles and cause clogging. Aggregation depends on various ink parameters, as discussed above.

b. Controlling the rheology of the ink by proper selection of the components which affect it most significantly — polymers (binder, dispersants) and the phase fraction of the dispersed particles. Since most inks for graphic applications contain particles below 10% v/v, the latter is less important. Therefore, the rheology control should be focused on the properties of the polymeric additives, such as molecular weight[17] and dissolution in the liquid components.

Drop Formation

Drop formation is probably the most important performance issue, since it affects the overall performance of the printing process. For example, if the jet disintegrates into many small droplets instead of one drop, the placement of the droplets will not be accurate, and the image quality will be very poor.

The jet stability and break-off behavior with respect to the fluid properties are stated in well-known theories such as Navier-Stokes equations and the Rayleigh theory.[18] During recent years many computer simulations have aimed at predicting the jetting process in specific print heads and, more importantly, for establishing a methodology for selection of ink additives.[2,19–21]

However, it appears that so far the actual selection of ink components is still based on an empirical approach, probably since the selection should take into consideration factors in addition to jetting, and since the exact chemical composition of most commercial additives (wetting agents, dispersants, etc.) is unknown.

Jetting results from pushing the liquid (by actuation of piezo-electric material or by bubble pressure) through the micron-size nozzles. The ejected fluid travels away from the nozzle with a specific momentum dictated by the kinetic energy of the droplet. Under optimal conditions, this fluid, which is actually an ink packet, possesses a detachment tail (ligament), which subsequently forms a drop due to surface tension effects, providing that the spacing between the nozzles and the substrate is large enough.[2] The main parameters which govern the jetting process are the surface tension and the rheological properties of the ink. Depending on the print head, typical inkjet inks should have a surface tension in the rage of 25–50 dyne/cm, and viscosity in the range of 1–25 cP. The relations between these parameters, together with ink density and nozzle diameter give a "window of operation" in terms of Reynolds numbers,[18,22,23] Weber numbers,[23] and the combined parameter which is the Ohnesorge number.[24] Quantitative information regarding these parameters is given in the chapter on water-based inks.

The ink formulator can control the surface tension by selection of the liquids of which the ink is composed (for example, addition of isopropanol to water causes a significant decrease in surface tension), or by adding surfactants which are effective at very low concentrations (usually silicon or fluoro surfactants). The use of surfactant may be problematic, since often it causes formation of bubbles, which may prevent proper jetting. Another issue related to the use of surfactants is dynamic surface tension[25]: since a new interface is formed very rapidly during jetting, the surfactant may not be present at the newly formed interface; thus the surface tension is higher than that measured at equilibrium. The diffusion of the surfactant to the interface takes time, depending on the structure of the surfactant, and results in a liquid jet having a surface tension that varies during the jetting and time of flight.

Ink viscosity can be controlled by the liquid compositions and the proper selection of the polymeric additives which may be present in the formulation (such as binders). After the initial ejection process, the ink ligament often breaks into a large drop and several small drops ("satellites"). The satellite issue is very complicated to control and leads to undesirable results on the printed substrate; since their velocity is different than that of the main droplet, they meet the substrate in a different position than that of the main drop. There are several reports which suggest minimizing satellite formation by changing the rheology of the ink.[26,27] Meyer et al.[27] demonstrated the effect of adding polyacrylamide to glycerol-water inks, and found that the break-off process can be controlled by changing the molecular weight (500 000–6 000 000) and the concentration of the polymer (10–200 ppm).

Materials Compatibility

The selection of materials for the ink should take into consideration the compatibility of each ink component with the construction materials it meets. These materials can be very different, starting from metallic parts (orifice plate, sensors, fittings, filters, etc.) which may degrade upon contact with a high or low pH medium and plastic parts which may dissolve or swell upon contact with organic solvent or UV monomers.

It should be emphasized that testing the compatibility of individual components of the ink is not sufficient, and compatibility (resistance) testing should be performed with the final ink. Usually such compatibility evaluations are performed by dipping the various parts of the print head and the ink supply system into the ink for a prolonged time under accelerated conditions, followed by analysis of the part for deterioration of its properties, performance, etc.

INK ON SUBSTRATES

The drop impact on the substrate converts the spherical drop into a flat dot, its size depending on the physicochemical properties of both substrate and ink. The behavior of ink drops on substrates depends on wetting substrates and the flow of ink during the wetting process. Three chapters in this book will deal with various aspects of wetting and ink-substrate interactions, and therefore only a short discussion regarding formulation guidelines will be given here.

a. *Spreading and wetting*: as a general rule, in order to obtain spontaneous wetting (large dot), the ink should have a surface tension lower than the surface energy of the substrate.[28,29] This means that, for example, for low energy surfaces such as polyethylene,[29,30] the surface tension should be below 28 erg/cm^2, if one wishes to obtain large dots. The use of low surface tension liquids (in solvent-based inks) to achieve that purpose is very limited, since most of such solvents are too volatile. Low surface tension values can be achieved by adding wetting agents, such as Byk 333®. It should be noted that since the spreading process usually takes place within 100–200 msec, the dynamic surface tension plays an important role here also. On the contrary, if small dots are required (for better resolution, increased height of the printed pattern in conductive inks, etc.), the surface tension should be high. Obviously, the surface tension of the ink should meet the requirements of the "jetting window parameters" as well.

b. *Flow*: The spreading of the drop is strongly dependent on its ability to flow on the substrate (for porous materials see the chapter on paper penetration of water-based inks; for rough surfaces see Ref. (31)). A high viscosity ink will lead to small dot size compared to a low viscosity ink, with a given time for the spreading process. The spreading usually stops when the viscosity becomes very high, for example after UV curing or solidification of hot melt ink. For both solvent- and water-based inks, the spreading may eventually stop due to evaporation of the liquids, especially on the thin film at the external part of the dot, which leads to increased viscosity. Therefore, prevention of drop spreading can be achieved by including volatile solvents in the ink that would cause a fast increase in viscosity upon evaporation.

It is interesting to note that a combined effect of surface tension gradients and evaporation may result in unique structures, namely ring structures. In the graphic arts this is usually a problematic phenomenon, but in functional printing it can be used to obtain interesting properties such as room temperature conductivity obtained by drops of conductive inks.[32–34]

SUMMARY

This chapter focused on information required to obtain inkjet inks with good performance. As discussed, obtaining good performance is based on meeting the requirements of each step of the printing process, including manufacture and storage. Some of the requirements in one step may contradict the requirements of another step, and therefore sometimes the optimal compositions are actually compromises between contradicting properties. For example, to have an ink with high latency and good recoverability would require an ink composed of solvents which never evaporate. However, such an ink will be useless once it is printed on a substrate, since there the ink should be present in its dry form.

It should be emphasized that there are other important issues to be taken into consideration while selecting ink compositions, in addition to the performance needs. The main issues are environmental, legislation, health and safety, flammability and last, but not least, ink cost.

REFERENCES

1. Calvert P. (2001) Inkjet printing for materials and devices. *Chem Mater* **13**: 3299–3305.
2. Clay K, Gardner I, Bresler E, Seal M, Speakman S. (2004) Direct legend printing (DLP) on printed circuit boards using piezoelectric inkjet technology. *Circuit World* **28**: 24–31.
3. Andreottola MA. (1991) Ink jet technology. In Diamond AS (ed.), *Handbook of Imaging Materials*, pp. 527–544. Marcel Dekker, New York.
4. Kang HR. (1991) Water-based ink-jet ink. II. Characterization. *J Imag Sci* **35**: 189–194.
5. Rosen MJ. (2004) *Surfactants and Interfacial Phenomena*, 3rd ed., pp. 332–339. Wiley Interscience, Hoboken, NJ.
6. Spinelli HJ. (1998) Polymeric dispersants in ink jet technology. *Advanced Materials* **10**: 1215–1218.
7. Wong R, Hair ML, Croucher MD. (1988) Sterically stabilized polymer colloids and their use as ink-jet inks. *J Imag Technol* **14**: 129–131.
8. Ink for ceramic Surfaces. European Patent Application No. 04770447.4.
9. Kang HR. (1991) Water-based ink-jet ink. I. Formulation. *J Imag Sci* **35**: 179–188.
10. Adamson AW. (1990) *Physical Chemistry of Surfaces*, 5th ed., pp. 4–49. Wiley Interscience, New York.
11. Nizri G, Lagerge S, Kamyshny A, Major DT, Magdassi S. (2008) Polymer-surfactant interactions: Binding mechanism of sodium dodecyl sulfate to poly(diallyldimethylammonium chloride). *J Colloid Interface Sci* **320**: 74–81.

12. Le HP. (1998) Progress and trends in ink-jet printing technology. *J Imag Sci Technol* **42**: 49–62.

13. Pond SF, Wnek WJ, Doll PF, Andreottola MA. (2000) Ink design. In Pond SF (ed.), *Inkjet Technology and Product Development Strategies*, pp. 153–204. Torrey Pines Research, Carlsbad, CA.

14. Ma Z, Anderson R. (2006) US Patent 7129284 B2.

15. Mendel J, Burner D, Bermel AD. (1999) Particle generation and ink particle size effects in pigmented inkjet inks — Part II. *J Nanopar Res* **1**: 421–422.

16. Bermel AD, Bugner DE. (1999) Particle size effects in pigmented ink jet inks. *J Imag Technol* **43**: 320–324.

17. Hiemenz PC. (1984) *Polymer Chemistry. The Basic Concepts.* Marcel Dekker, New York.

18. Kang HR. (1991) Water-based ink-jet ink. III. Performance studies. *J Imag Sci* **35**: 195–201.

19. Chen PH, Chen WC, Ding PP, Chang SH. (1998) Droplet formation of a thermal sideshooter inkjet printhead. *Int J Heat Fluid Flow* **19**: 382–390.

20. Wu HC, Lin HJ, Kuo YC, Hwang WS. (2004) Simulation of droplet ejection for a piezoelectric inkjet printing device. *Mater Transact* **45**: 893–899.

21. Okumura K, Ogawa H, Endo Y, Kaiba T. (2002) Japan Patent 2002294540 A.

22. Tajiri K, Suzuki K, Nakamura T, Muto K. (2007) Japan Patent 2007063087.

23. Noguera R, Dossou-Yovo C, Lejeune M, Chartier T. (2005) 3D fine scale PZT skeletons of 1–3 ceramic polymer composites formed by ink-jet prototyping process. *Journal de Physique IV: Proceedings* **126**: 133–137.

24. Girard F, Attane P, Morin V. (2006) A new analytical model for impact and spreading of one drop: Application to inkjet printing. *Tappi Journal* **5**: 24–32.

25. Pistagna A, Gino L. (2007) US Patent 7160373.

26. Lee S, Webster GA, Pietrzyk JR, Barmaki F. (2004) US Patent 6790268 B.

27. Meyer J, Bazilevsky AV, Rozhkov AN. (1997) Effect of polymeric additives on thermal ink jets. *Proceeding of the International Conference on Digital Printing Technologies*: 675–680.

28. Rosen MJ. (2004) *Surfactants and Interfacial Phenomena,* 3rd ed., pp. 243–276. Wiley Interscience, Hoboken, NJ.

29. Podhajny RM. (1991) Surface tension effects on the adhesion on the drying of water-based inks and coatings. In Sharma MK & Micale FJ (eds.), *Surface Phenomena and Fine Particles in Water-Based Coatings and Printing Technology,* pp. 41–58. Plenum Press, New York.

30. He D, Reneker DH, Mattice WL. (1997) Fully atomistic models of the surface of amorphous polyethylene. *Comput Theor Polym Sci* **7**: 19–24.

31. Apel-Paz M, Marmur A. (1999) Spreading of liquids on rough surfaces. *Colloids Surf A Physicochem Eng Asp* **146**: 273–279.

32. de Gans DJ, Schubert US. (2004) Inkjet printing of well-defined polymer dots and arrays. *Langmuir* **20**: 7789–7793.

33. Magdassi S, Grouchko M, Toker D, Kamyshny A, Balberg I, Millo O. (2005) Ring stain effect at room temperature in silver nanoparticles yields high electrical conductivity. *Langmuir* **21**: 10264–10267.

34. Xia Y, Friend RH. (2007) Nonlithographic patterning through inkjet printing via holes. *Appl Phys Lett* **90**: 253513/1–253513/3.

CHAPTER 3

Equilibrium Wetting Fundamentals

Abraham Marmur
*Department of Chemical Engineering,
Technion — Israel Institute of Technology,
32000 Haifa, Israel*

INTRODUCTION

Printing with ink is about wetting. It starts with the first contact of ink and the substrate, goes through a dynamic phase of spreading of ink over (or inside) the substrate, and ends with the ink getting to equilibrium with the substrate and the environment.[1-3] All of these processes are very complicated, and strongly depend on the nature of the ink, the nature of the substrate, and the environmental conditions. The simplest and most fundamental process underlying all of them is wetting of a non-porous, solid substrate by a drop of a "simple" liquid (e.g., nonvolatile, and without surfactants). This fundamental process may appear to be somewhat remote from the real one; however, its understanding is a prerequisite for mastering the more complicated processes that occur in realistic printing situations.

Out of the abovementioned three stages, the equilibrium stage is the one that is most critical for print quality. Therefore, this chapter focuses on equilibrium wetting. Two main aspects will be discussed: (a) the conditions that determine the extent of wetting, and (b) the experimental approach to the characterization of the wettability of a solid substrate. Obviously, they are closely linked, and originate from the same equations. Equilibrium wetting is characterized, theoretically and practically, by the contact angle (CA), which is defined as the angle between the tangent to the liquid-air interface and the tangent to the solid-liquid interface. Since most solid surfaces may be rough or chemically heterogeneous, a distinction must be made between the apparent contact angle (APCA) and the actual contact angle (ACCA). The former is defined by a macroscopic measurement, which ignores the fine details of the solid surface; the latter is conceptually defined at the actual contact line between the liquid and the solid, and cannot be experimentally accessed by currently available techniques. Figure 1 demonstrates the APCA and ACCA in a simplistic way.

The following discussion starts with the concept of equilibrium wetting on an ideal solid, which is defined as smooth, rigid, chemically homogeneous, insoluble, and non-reactive. The formal theory of wetting on ideal surfaces started about two hundred years ago.[4] However, mathematical difficulties have prevented so far a complete understanding of wetting on real surfaces with three-dimensional topographies. Nonetheless, much progress has been done over the years with regard to two-dimensional (e.g., axisymmetric) situations, and some analytical and numerical advancement has been recently achieved concerning three-dimensional cases.[5] These developments will also be described below, and will serve as a basis for understanding wetting behavior as well as wetting characterization.

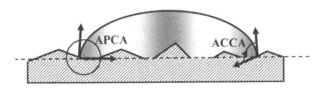

Fig. 1. A simplistic view of drop on a rough solid surface, demonstrating the difference between the actual contact angle (ACCA) and the apparent one (APCA).

WETTING OF IDEAL SURFACES

The ACCA and APCA on an ideal solid surface are identical by definition, and are referred to as the "ideal contact angle (ICA)". As will be explained below, it is the value of the ICA that is required for the characterization of the wettability of a solid substrate in terms of its surface tension. Also, all predictions of wetting behavior start with the ICA as their basis. Therefore, even though ideal surfaces are rarely encountered in practice, the concept is of fundamental importance.

The pioneering equation for the ICA is based on the Young theory[4]

$$\cos \theta_Y = \frac{\sigma_{sf} - \sigma_{sl}}{\sigma_{lf}} \tag{1}$$

where θ_Y is the Young contact angle (YCA), and σ_{lf}, σ_{sl}, and σ_{sf} are the liquid-fluid, solid-liquid, and solid-fluid interfacial tensions. With regard to printing applications, the fluid is usually air; therefore it will be assumed throughout the rest of this chapter to be air (or, in general, a vapor). Since the density of a vapor is much lower than that of a liquid or a solid, it hardly affects interfacial tensions. Therefore, σ_{sf} will be replaced below by σ_s (the surface tension of the solid), and σ_{lf} by σ_l (the surface tension of the liquid). As shown by Eq. (1), the YCA depends only on the physicochemical properties of the three phases. Gravity, as an example of an external force, affects the shape of the liquid-vapor interface, but not the CA itself. Thermodynamically, the YCA represents the state of minimal Gibbs energy of the solid-liquid-vapor system. It is important to realize that for an ideal solid surface there is only a single minimum in the Gibbs energy. Therefore, the CA has a single value on an ideal surface.

Out of the parameters appearing in the Young equation, Eq. (1), only the surface tension of the liquid can be simply and accurately measured. However, the measured value of σ_l can be used in the Young equation only under a certain condition, as discussed in the following. Gibbs was the first to note that the Young equation may need to be modified, even for an ideal solid surface. This is so because the three interfacial tensions may be influenced by each other at the three-phase contact line, due to the effect that one phase may have on the interaction between the other two phases. In the present case, the main effect would be that of the solid on the value of σ_l, very close to the contact line. In other words, the value of σ_l in the Young equation may not be the same as the experimentally measured far away from the solid. Gibbs suggested accounting for this phenomenon by a "line tension", an approach which is conceptually similar to the use of surface tension to account for two-phase interactions. Thus, the measured value of σ_l may be used in Eq. (1) only if the line-tension correction is negligible. An order of magnitude analysis indeed showed that, for all practical purposes, line tension effect on the YCA is negligible for macroscopic drops.[6]

The other interfacial tensions appearing in the Young equation, σ_s and σ_{sl}, are not accessible to measurement as yet. The former is of special interest, because it is the solid surface characteristic of interest. In addition, the YCA can be measured only in very few cases that involve solid surfaces, which are close to being ideal. Thus, even if the values of θ_Y and σ_l are known, Eq. (1) still contains two unknowns. In order to elucidate σ_s from the Young equation, or predict wetting behavior on a predefined solid surface, the value of σ_{sl} must be known.

This information may be provided by various semi-empirical correlations that have been developed over the years from the equations of Girifalco and Good,[7] and Fowkes.[8] For example, such a correlation can be

$$\sigma_{sl} = \sigma_s + \sigma_l - 2\phi\left(\sigma_s, \sigma_l\right)\sqrt{\sigma_s\sigma_l} \tag{2}$$

where ϕ is a function of the surface tensions. Owens and Wendt,[9] for example, assumed surface tension to consist of a non-polar (dispersive) contribution and a polar one:

$$\sigma = \sigma^d + \sigma^p \tag{3}$$

where the superscript d indicates the dispersion (van der Waals) contribution, and the superscript p stands for the polar contribution. In this case, the function ϕ is given by

$$\phi = \frac{\sqrt{\sigma_s^d\sigma_l^d} + \sqrt{\sigma_s^p\sigma_l^p}}{\sqrt{\sigma_s\sigma_l}} \tag{4}$$

Theoretical and experimental estimates show that the value of ϕ is between about 0.55 and 1.2. Additional correlations are discussed elsewhere in the literature.[10]

The surface tension of the solid, σ_s, can be calculated (together with σ_{sl}) from the set of Eqs. (1) and (2). To accomplish this calculation, the value of the function ϕ must be known; however, this is seldom the case. Actually, the number of unknowns is usually larger than two, because of the various components that make up the surface tension (such as σ^d or σ^p) are unknown. To solve this problem, additional CA measurements are made, using a few different liquids, but the results may strongly depend on the choice of the liquids used.[11] Moreover, the choice of the proper correlation is still an open challenge. However, in principle, the surface tension of a solid surface and its YCA can be correlated with each other.

WETTING OF REAL SURFACES

As mentioned in the introduction, most surfaces encountered in practice are to some extent rough or chemically heterogeneous, or both. Such surfaces, in contrast to the concept of an ideal surface, may be referred to as real surfaces. It was rigorously proven that the ACCA is the YCA, if the effect of line tension can be neglected.[12] However, the ACCA is not yet accessible to experimental measurement; therefore, the YCA is also not experimentally accessible. In contrast, the APCA can be experimentally measured, since, by definition, it is a macroscopic characteristic of wetting. Thus, one of the main challenges of the theory of wetting is to explain the relationship between the APCA and the YCA. Understanding the correlation between these two angles enables the following two complementary accomplishments: (a) prediction of macroscopic wetting behavior, expressed in terms of the APCA, from the value of the YCA, and (b) characterization of a solid surface in terms of the YCA (from which σ_s can be calculated), based on measurements of the APCA.

The definition of the APCA in a two-dimensional picture (e.g., Fig. 1) is obvious. However, in three-dimensional situations it needs further clarification. It was shown, theoretically and numerically, that a drop becomes more axisymmetric as it gets bigger with respect to the scale of chemical heterogeneity or roughness.[13,14] When the drop becomes axisymmetric, all shape distortions due to heterogeneity or roughness turn out to be limited to the close proximity of the contact line. Therefore, a macroscopic observation of the drop picks up an axisymmetric shape with a clearly defined APCA. Thus, the APCA of a drop on a real surface is meaningful and can be properly measured only for sufficiently large drops. The following discussion is, therefore, limited to such large drops.

The equilibrium condition of a wetting system at constant pressure and temperature can be expressed by a minimum in the Gibbs energy. For a drop on an ideal solid surface there is only a single minimum in the Gibbs energy, at the APCA that is given by the Young equation, Eq. (1) (in this case the APCA is equal to the ACCA). However, for a drop on a real surface, the curve of Gibbs energy vs. APCA exhibits multiple minima (see Fig. 2). The reason for this multiplicity is that the fundamental equilibrium condition,

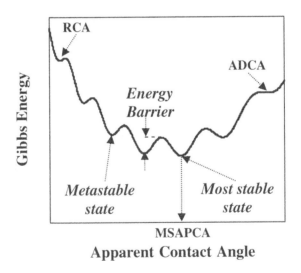

Fig. 2. A simplified drawing of the Gibbs energy curve for a real wetting system. Each minimum defines a metastable state. The lowest minimum defines the most stable state and the most stable, apparent contact angle (MSAPCA). The lowest and highest APCAs that are associated with a metastable state are the receding contact angle (RCA) and the advancing contact angle (ADCA), respectively.

namely the equality of the ACCA and the ICA,[12] is fulfilled at more than one location. Consequently, different values of the APCAs may be associated with the various metastable equilibrium states. The one with the lowest energy is the most stable state, and the APCA associated with it is referred to as the most stable apparent contact angle (MSAPCA). The other equilibrium states are metastable. A drop can move from one metastable state to a lower energy one if external energy is supplied to overcome the energy barrier between them, e.g., via vibrations. It is of interest to notice that the energy barriers are the highest near the most stable state.

Looking at the Gibbs energy curve it becomes clear that it depicts three distinctive CAs: the lowest APCA for which there is a local minimum, the MSAPCA, and the highest APCA associated with a local minimum. The latter is called the advancing contact angle (ADCA), and the lowest metastable APCA is the receding contact angle (RCA). The difference between the ADCA and the RCA is called the hysteresis range. The question that needs to be discussed is: which of these three CAs is most useful in terms of behavior prediction or characterization? The ADCA and RCA are experimentally measurable; however, there is no substantiated theory so far that correlates them with the YCA. In addition, the measured ADCA and RCA may be different from the theoretical ones, since natural vibrations in the laboratory may prevent the system from reaching the theoretical values (the energy barriers near the ADCA or the RCA are small). On the other hand, it is more difficult to experimentally identify the MSAPCA, but the theory behind it is relatively well substantiated, as described in the following. Thus, the question posed above has no unique answer so far; it may depend on the practical application of interest, and on availability of data. Obviously, further theoretical understanding is required for a more definite answer.

In order to theoretically predict the MSAPCA, it is necessary to pinpoint the lowest minimum in the Gibbs energy curve. Mathematically, to identify the lowest minimum among multiple minima is possible only by checking all minima one by one. Therefore, no analytical way exists to immediately recognize the most stable equilibrium state of a wetting system. Nevertheless, a useful approach was suggested many years ago: to presume that the solid surface is ideal, with surface properties that are the average of the actual properties over the whole surface. This approach was used for rough surfaces,[15] as well as for chemically heterogeneous ones.[16]

For a rough solid surface, the area is bigger than its projected area by a factor r, the roughness ratio. If the surface is chemically homogeneous, this increase in the solid-liquid and solid-fluid contact areas can be alternatively considered as an increase in the interfacial tensions, σ_{sl} and σ_s, by the same factor r. This equivalence is true, since each term contributing to the Gibbs energy of a wetting system consists of a product of the corresponding interfacial tension by the contact area; thus, mathematically it does not make any difference which of the elements of the products is multiplied by the factor r. This is the consideration underlying the Wenzel equation[15]

$$\cos \theta_W = r \cos \theta_Y \tag{5}$$

where the subscript W indicates the MSAPCA calculated according to this equation. As mentioned above, the discussion of APCAs is relevant only for sufficiently large drops. It turns out that the Wenzel equation may serve as a good approximation for the MSAPCA also only when the drop is sufficiently large compared with the roughness scale.[13] The practical question of how large should the drop be has not yet been answered.

For a chemically heterogeneous surface the same averaging approach directly yields the Cassie equation[16]

$$\cos \theta_C = \sum_i x_i \cos \theta_{Yi} \tag{6}$$

In this equation, θ_C is the MSAPCA, x_i is the area fraction of the solid surface that is characterized by a given chemistry i, and θ_{Yi} is the YCA for each chemistry i, had it covered the whole surface. This form generalizes the original Cassie equation that was based only on two chemistries. The summation can be done in a form of an integral when the chemical heterogeneity is continuously varying over the surface. As demonstrated by numerical simulations,[14] the Cassie equation also becomes a good approximation when the drop is sufficiently large compared with the wavelength of chemical heterogeneity.

Thus, using the Wenzel and Cassie equations, one can correlate the YCA with the MSAPCA. For a rough, but chemically homogeneous surface, one can predict what the MSAPCA would be if the YCA is known (e.g., by measuring the CA on a smooth surface of the same material) and r is measured. Interestingly, $\theta_W < \theta_Y$ if $\theta_Y < 90°$, and $\theta_W > \theta_Y$ if $\theta_Y > 90°$. This observation may serve as an important way of controlling wettability by the roughness of the surface. Conversely, if the MSAPCA and r are measured, the YCA can be calculated. For chemically heterogeneous surfaces, the situation is somewhat more complicated. Indeed the MSAPCA may be calculated from the Cassie equation if all the YCAs of the various chemistries are known. However, going back from the average (i.e. θ_C) to find out the individual APCAs seems impossible at this stage.

To directly measure the MSAPCA a working hypothesis must be assumed, namely that the system can be brought to its most stable energy state by applying vibrations.[17] This approach has been tried for a few systems, with promising results. However, since the ADCA and RCA may be more amenable to direct experimental measurement, it would be useful to be able to calculate the MSAPCA from them. As mentioned above, there is no theory yet that supports such a calculation. However, some empirical suggestions were made, based on a limited number of experiments. One suggestion is[18]

$$\cos \theta_{ms} = (\cos \theta_a + \cos \theta_r)/2 \qquad (7)$$

where θ_{ms} is the MSAPCA, θ_a is the ADCA, and θ_r is the RCA. Another correlation[19] suggests a different average

$$\theta_{ms} = (\theta_a + \theta_r)/2 \qquad (8)$$

Hopefully, future theory will be developed to settle this question. In the meantime, it may be important to take note of some observations regarding the measurement of the ADCA and RCA. One observation is that the ADCA and RCA may depend on the drop volume.[20] This dependence seems to diminish at large drop volumes. Thus, it is very important to be aware of this dependence when measuring the

ADCA and RCA, and use large drop volumes. Another important observation is that contact angle hysteresis appears to depend on the type of the system. For example, the hysteretic behavior of the CA when using the captive bubble measurement method (an air bubble under the liquid) will be different than when using the sessile drop method (a liquid drop in air).[20] It is possible to take advantage of this observation by choosing the measurement method in a way that minimizes the experimental uncertainty in the measured CA for a given solid surface. It turns out that the oscillations in the APCA are smaller when using the captive bubble method for low-energy surfaces than when using the drop method, and vice versa. Smaller oscillations reduce the experimental uncertainty in determining the ADCA and RCA.

SUMMARY AND CONCLUSIONS

The fundamental theoretical questions underlying the wetting of a solid surface by a liquid drop have been described and discussed. These theoretical principles can be directly applied to practice along two main lines: (a) characterization of solid surfaces in terms of their surface tension; and (b) designing processes based on controlling wettability properties. The following points summarize current understanding for each of these two directions.

Assessment of Surface Tension of the Solid

- The most stable contact angle should preferably be measured; alternatively, the advancing and receding contact angles can be measured and the most stable one can be estimated from them (Eq. (7) or (8) or, hopefully, a more substantiated future theory).
- For rough surfaces, the Young contact angle should be calculated from the most stable contact angle by the Wenzel equation (Eq. (5)), using the value of the roughness ratio (independently measured or calculated); for chemically heterogeneous surfaces, the most stable contact angle is the average Young contact angle of the surface.

- From the Young contact angle the surface tension is assessed, using Eq. (1) and a correlation for interfacial tensions, such as given by Eq. (2); some correlations require repetition of the process with a few different liquids in order to be able to assess the surface tension.

Designing Wetting Processes

- The Young contact angle is the starting point for any design.
- Wetting may be enhanced by roughness if the Young contact angle is smaller than 90° (the most stable contact angle is smaller than the Young one, and may lead to complete wicking); wetting is hampered by roughness if the Young contact angle is higher than 90°.
- Wetting may be strongly affected by chemical heterogeneity; it is determined by the average Young contact angle, which is calculated by weighing the area fraction of each chemistry.

REFERENCES

1. Dussan V EB. (1979) Spreading of liquids on solid surfaces — static and dynamic contact lines. *Annu Rev Fluid Mech* **11**: 371–400.
2. Marmur A. (1983) Equilibrium and spreading of liquids on solid surfaces. *Adv Coll Int Sci* **19**: 75–102.
3. DeGennes PG. (1985) Wetting: statics and dynamics. *Rev Mod Phys* **7**: 827–863.
4. Young T. (1805) An essay on the cohesion of fluids. *Philos Trans R Soc (Lond)* **95**: 65–87.
5. Marmur A. (2006) Soft contact: measurement and interpretation of contact angles. *Soft Matter* **2**: 12–17.
6. Marmur A, Krasovitski B. (2002) Line tension on curved surfaces: liquid drops on solid micro- and nano-spheres. *Langmuir* **18**: 8919–8923.
7. Girifalco LA, Good RJ. (1957) A theory for the estimation of surface and interfacial energies. I. Derivation and application to interfacial tension. *J Phys Chem* **61**: 904–909.

8. Fowkes FM. (1963). Additivity of intermolecular forces at interfaces. I. Determination of the contribution to surface and interfacial tensions of dispersion forces in various liquids. *J Phys Chem* **67**: 2538–2541.
9. Owens DK, Wendt RC. (1969). Estimation of the surface free energy of polymers.*J Appl Polym Sci* **13**: 1741–1747.
10. Van Oss CJ, Good RJ, Busscher RJ. (1990). Estimation of the polar surface tension parameters of glycerol and formamide, for use in contact angle measurements on polar solids. *J Dispers Sci Technol* **11**: 75–81. ·
11. Shalel-Levanon S, Marmur A. (2003) Validity and accuracy in evaluating surface tension of solids by additive approaches. *J Coll Int Sci* **262**: 489–499; **268**: 272.
12. Wolansky G, Marmur A. (1998) The actual contact angle on a heterogeneous rough surface in three dimensions. *Langmuir* **14**: 5292–5297.
13. Wolansky G, Marmur A. (1999) Apparent contact angles on rough surfaces: the wenzel equation revisited. *Colloids Surf A* **156**: 381–388.
14. Brandon S, Haimovich N, Yeger E, Marmur A. (2003) Partial wetting of chemically patterned surfaces: the effect of drop size. *J Coll Int Sci* **263**: 237–243.
15. Wenzel RN. (1936) Resistance of solid surfaces to wetting by water. *J Ind Eng Chem Res* **28**: 988–994.
16. Cassie ABD. (1948). Contact angles. *Discuss Faraday Soc* **3**: 11–16.
17. Meiron TS, Marmur A, Saguy IS. (2004) Contact angle measurement on rough surfaces. *J Coll Int Sci* **274**: 637–644.
18. Andrieu C, Sykes C, Brochard F. (1994). Average spreading parameter on heterogeneous Surfaces. *Langmuir* **10**: 2077–2080.
19. Decker EL, Garoff S. (1996). Using vibrational noise to probe energy barriers producing contact angle hysteresis.*Langmuir* **12**: 2100–2110.
20. Marmur A. (1998) Contact angle hysteresis on heterogeneous smooth surfaces: theoretical comparison of the captive bubble and drop methods. *Colloids Surf A* **136**: 209–215.

CHAPTER 4

The Behaviour of a Droplet on the Substrate

Patrick J. Smith

The University of Freiburg, Baden-Württemberg, Germany

INTRODUCTION

The behaviour of droplets, as a whole, is a complex area of interest to researchers in a variety of fields, such as combustion, aerosol science and spray technology, to mention a few. Recently, Frohn and Roth[1] have written about many aspects of droplet dynamics and the reader is referred to their book as a good introduction to the subject. However, from an inkjet point of view the researcher is primarily interested in the behaviour of droplets on the substrate, from the moment when the droplet lands to the droplet's end-state. In the first instance, the researcher strives to ensure that droplets land in their specified locations and remain there. Then, they try to control the droplets' behaviour in order to obtain the desired morphology. The aim of this chapter is to discuss the lifetime of inkjet-produced droplets from the moment of their impact with the substrate to the moment when the final feature is obtained, which is typically by evaporation of the droplet's carrier solvent or by the solidification of said solvent.

Where possible, this chapter focuses on droplets produced by inkjet printing, which are usually picolitre-sized. However, inkjet researchers often cite a number of papers that are based on the observations of microlitre-sized droplets, which have been dispensed by pipette, and several of these papers will also be covered since their findings are often used to describe inkjet printed droplets. Typically, single nozzle systems have been used to produce the inkjet printed droplets, with treated glass slides being the usual substrate.

One can either control droplets on the substrate by chemically confining them in pre-prepared regions of preferential wetting or mechanically, by creating embossed features. However, most researchers investigating the inkjet printing of experimental inks typically use fairly smooth, glass slides partly on account of their low cost and ready availability. Also, as one of the oft-spoken advantages of inkjet printing is its ability to produce patterns without the need for masks, it can seem somewhat paradoxical to involve an additional patterning step.

Therefore, the following discussion aims to explain how droplets behave on smooth, unstructured substrates, which have been either coated or cleaned, and why certain phenomena might occur. After discussing droplet impact, two types of droplet will be considered. The first are droplets of solvent, which may also describe droplets of solution below the solution's critical concentration. The second are suspension droplets that can describe droplets of solution after the critical concentration has been exceeded and are useful in explaining droplets formed from nanoparticle ink. Besides single droplets, large arrays of droplets, such as lines and films, will also be discussed.

There are several types of substrate that could be considered. However, substrates that have been patterned by micro- or nano-structures or are porous, such as paper or textiles are beyond the scope of this chapter. The two main types of substrate that have been used are either smooth with a high surface energy, namely a low contact angle which is less than $90°$, typically $< 40°$, or have similar smoothness but with a low surface energy, i.e. a contact angle that is greater than $90°$. (The value of all contact angles given will be in degrees since this value is more readily visualised.) A few experiments have been performed on relatively rough substrates and these will be discussed where relevant.

The final sections of the chapter will first discuss how droplet behaviour on unstructured substrates can be controlled before looking at inkjet printed films and lines, and how to obtain uniform morphology in these features. Finally, a brief summary of the chapter as a whole is given.

Droplet Impact

The process of a droplet with finite velocity hitting a solid surface can be divided into three stages. In the first stage, the droplet makes contact with the substrate. In the second, a thin, circular film forms around the droplet (the shape of the film differs if the droplet's angle of approach is slanted towards the substrate). The radius of the circular film expands to an order of magnitude greater than that of the in-flight droplet radius. As the film expands, liquid travels radially outwards and mass is accumulated at the boundary forming a liquid ring. Kinetic energy is dissipated partly due to viscous flow in the thin film.

In the final stage, the film begins to recoil when a maximum is reached, the recoil may, in some situations, lead to the droplet actually detaching from the surface. The fluid comes to rest after a series of inertial oscillations that are dampened by viscous dissipation. Obviously, the dynamic contact angle changes from the advancing angle to the receding angle during recoil.

Splashing can be seen to begin in droplets with Weber (We) numbers of 100–1000, and fingers have been observed in droplets that have a Reynolds (Re) number of 15 000 and a We of 1000. (Re $= (\rho.u.d)/\mu$ and We $= (\rho.u^2.d)/\sigma$, where ρ, u, d, μ and σ are the liquid density, droplet impact velocity, droplet diameter, liquid viscosity, and liquid surface tension respectively.) As the drops used in inkjet printing typically have diameters below 100 microns, values of Re $= 2.5–2000$, and We $= 2.7–1000$ can be expected.

Van Dam and Le Clerc have observed, using a CCD camera with a microscope objective, the impact of water droplets[2] on glass slides, which had been thoroughly cleaned and treated to produce advancing contact angles of either 15°, 35° or 70°. By using inkjet printing, they were able to produce droplets with diameters in the range of 36–84 μm. In 26 out of 28 experiments, they observed a small air bubble in the inkjet printed droplet. Smaller bubbles were seen when high impact velocities and large Weber numbers were used. They determined that the surface energy of the substrate did not influence the formation of the bubble. Instead, they attributed its formation to the air that is trapped between the falling droplet and the underlying substrate.

The image shown in Fig. 1 illustrates the impact of a water droplet with an initial diameter of 85 μm, a velocity of 5.1 m/s and a Weber number of 30. The three stages of impaction discussed earlier can be seen. Van Dam and Le Clerc observed that the average contact angle at the moment the inertial oscillations stopped was significantly smaller than the average advancing contact angle, but still larger than the receding contact angle, when the 70° substrate and a large Weber number were used.

During the first phase of impact, the droplet deforms and flattens when the We is less than 1.1, whereas it remains relatively unchanged for higher values. For high-speed impacts, the time scale for spreading becomes considerably smaller than the time scale for deformation of the droplet by surface tension.

In the second stage of impact, a liquid ring is seen to form. The thickness of this ring is not affected by surface energy but does vary with We. For example, when We was less than 2.4 no thickness was seen, whereas when We was greater than 24, the liquid ring was very thick, as can be seen in the second row of Fig. 1. When droplets landed with relatively high velocities (e.g. 10 m/s and larger), a thin sheet of fluid is seen to form along the substrate. This is known as a "wall jet".

Van Dam and Le Clerc also measured the droplet's final radius on the substrate and used this as a comparison for values predicted by several models. They found that the model of Pasandideh-Fard et al.,[3] gave the most accurate predictions: $R_f/R_o = ([We+12]/[3(f_s - \cos\theta_d)+4(We/Re^{-2})])^{-2}$, where R_f = final radius, R_o = initial radius, f_s = the ratio of fluid-vapour surface to fluid-solid surface, which can be taken as 1 for small contact angles, and θ_d = the contact angle

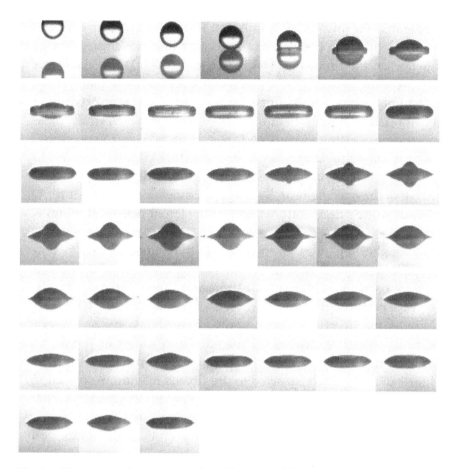

Fig. 1. The impact of a water droplet with an initial diameter of 85 μm, a velocity of 5.1 m/s and a Weber number of 30. The contact angle formed by water with the substrate was 35°. The delay between each picture is 3 μs, with whole sequence taking 135 μs. (From Ref. 2 © 2004 American Institute of Physics.)

formed by the droplet when it reaches its maximum radial extent. An impacting droplet tends to reach its final radius just after the second stage is ended but before the inertial oscillations begin. The increase in radius and the radius's final value is greater for larger values of We.

The behaviour of molten droplets landing on solid substrates that are composed of the same material at lower temperature has been investigated for low Weber numbers.[4] The spread of such droplets is controlled by the freezing of the contact line, which has a large influence on the final solid shape of the droplet. When the We is low, spreading is driven by interfacial forces that take place at the contact line rather than by impact phenomena. For low We droplet deposition, the Ohnesorge number ($Z = \mu/(\rho \cdot \sigma \cdot d)^{-2}$), which measures the ratio of the viscous and inertial resistances to spreading, is of use; its value decreases with increasing droplet size. For a droplet that is solidifying, the Stefan number ($S = (c[T_f - T_t]/L$, where c is the specific heat, T_f is the solidification temperature, T_t is the substrate's temperature and L is the latent heat of melting) is important.

Although they used droplets with diameters of 2 mm and more, the work of Park et al.[5] is interesting on account of the fact that they used four different substrates and four different liquids. They observed the impact of droplets of distilled water, n-Octane, n-Tetradecane or n-Hexadecane onto glass slides, silicon wafers, HMDS (Hexamethyl disilazane) coated silicon wafers or Teflon, for Reynolds numbers from 180 to 5513 and Weber numbers from 0.2 to 176. A model was constructed to predict the maximum spreading ratio, which is the ratio of the maximum spreading diameter to the initial droplet's diameter, for low impact velocities.

Water droplets of 2.3 mm diameter were observed landing on each of the four substrates at room temperature. The impact velocity was 1 m/s and the Re and We numbers were calculated to be 2000 and 24 respectively. It was found that the maximum spreading ratio was reached after 1.7 ms for the two substrates with the lowest surface energy, i.e. Teflon and HMDS silicon. Whereas, the maximum spreading ratios on the glass slide and the silicon wafer are only about 10–20% greater, taking 2.0 ms to be reached. When impact speeds were increased to 2.36 m/s the droplet was seen to rise from the Teflon substrate during the recoil. The maximum spreading ratio increases with increasing impact for all substrates, as one would expect.

The impacting process for high surface energy substrates has two stages. First, the droplet spreads to the maximum spreading ratio and then it retracts to the equilibrium spreading ratio. The impacting process for a droplet on low surface energy substrates, however, includes spreading, retracting and sometimes rebounding if the impact velocity is high enough, and then spreading to the equilibrium spreading ratio. The final equilibrium spreading ratio is primarily determined by the solid-liquid interaction and varies slightly with impacting velocity.

The impacting and spreading process is dominated by kinetic energy. For impact velocities greater than zero, there is an initial period where the spreading is driven by the kinetic energy of the impacting drop, and the spreading ratio increases much faster. Energy dissipation includes both viscous dissipation in the bulk fluid and dissipation in the vicinity of the contact line. The influence of viscous dissipation was observed for 2 mm diameter droplets of n-Octane and n-Hexadecane that were deposited on a Teflon film surface with an impact velocity of 1.9 m/s. Although the Weber numbers were similar (158 for n-Octane and 160 for n-Hexadecane), the Reynolds numbers differed due to n-Hexadecane being about 6.2 times more viscous than n-Octane. Re was 4087 for n-Octane and 856 for n-Hexadecane. The viscous effects relative to inertial effects are about 4.8 times greater for n-Hexadecane. The spreading ratios of both were similar during the early stages of spreading. However, the maximum spreading ratio was about 77% lower for the n-Hexadecane droplet than for the n-Octane droplet due the higher viscous dissipation in the n-Hexadecane droplet. Although the n-Octane drop has a greater distance to cover, both liquids take about the same time to retract to the resting position.

Evaporation of a Sessile Droplet of Solvent at Room Temperature

The influence of evaporation on the contact angle of a sessile droplet of solvent has been studied.[6] There are three regimes that a solvent droplet on a smooth substrate undergoes during evaporation. In the first regime, the diameter of the droplet remains constant; alternatively one can say that the droplet-substrate area is unaffected. As the droplet loses mass, through evaporation, the contact angle and the droplet's height decrease until the receding angle is reached. Thereupon, the second regime begins in which the droplet's diameter begins to shrink. The contact angle remains constant while the droplet-substrate area diminishes. Finally, both the droplet's diameter and contact angle shrink until evaporation is complete. As a general rule, droplets with smaller initial contact angles have higher evaporation rates and mass loss is linear.

The evaporation of a droplet on a rough surface behaved similarly except that the second regime was not entered; the diameter stayed constant due to the pinning of the droplet's contact line by the rough surface.

The substrate also has a role to play in the evaporation of a droplet,[7] with the evaporation rate of droplets being limited by the substrate's thermal properties; this is especially the case for high evaporation rates. When a droplet is deposited onto a substrate, two extreme cases can occur. If the substrate is a perfect thermal insulator then the evaporation rate is altered by changes in the droplet-vapour interface area. However, if the substrate is a perfect thermal conductor then the evaporation rate is also affected by a second mechanism, namely the heat transfer between the substrate and the droplet. In this second situation, the evaporation rate is higher than that of a droplet sitting on a thermal insulator.

The differences in temperature measured in the centre of a droplet and the temperature of its atmosphere are shown in Table 1 for three solvents. Higher differences were seen for the droplets that were placed upon the insulating substrates.

Table 1. Temperature difference between ambient air and droplet bulk. (From Ref. 7 © 2006 Elsevier B. V.)

	Water	Methanol	Acetone
Al	0 K	3.6 K	4 K
Ti	0 K	—	—
Macor	0.6 K	—	—
PTFE	1.3 K	6.9 K	8.6 K

It should be noted that glass, which is often used as a substrate in inkjet printing experiments, is an insulator. Similarly, although thermo-capillary surface instabilities have been seen to develop in the spreading and evaporation of volatile drops on conductive substrates (silicon and brass), no surface oscillations were observed for those droplets placed on glass.[8]

Evaporation of a Sessile Droplet of Suspension at Room Temperature

An evaporating, sessile droplet of suspension on a smooth substrate can be thought of as behaving similarly to the solvent droplet on a rough substrate that was described in the previous section. In the initial stage of evaporation, the droplet's diameter remains constant while the droplet's contact angle and height decrease. However, as the droplet is composed of suspended particles (or precipitate in the case of a solution droplet) and carrier solvent, its diameter remains constant because some of these particles are deposited close to the contact line, which pin the droplet.[9] As evaporation continues, a replenishing flow of solvent, carrying suspended material with it, travels from the droplet's centre to its pinned edge. This process continues until evaporation is complete and results in a feature that is commonly called a coffee stain (Fig. 2).

Deegan *et al.*,[10] gave three conditions that need to be met by the droplet in order for coffee staining to occur. These conditions are that the solvent meets the substrate at a non-zero contact angle, the contact line is pinned and that the solvent is volatile. These conditions were subsequently refined to being a pinned contact line and evaporation from the edge of the droplet.[9]

The actual conditions that are the cause of coffee staining continue to generate debate; mostly centred on the question of whether the contact line is initially pinned. Deegan used droplets of a colloidal suspension on atomically smooth mica to explore the issue of self-pinning.[11] He said that rings form because the contact line cannot move and that some pre-existing conditions on the substrate temporarily anchor it until solid material accumulates. However,

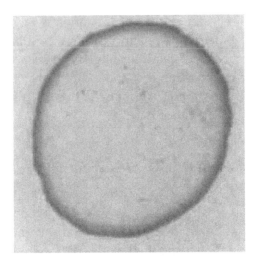

Fig. 2. A dried coffee droplet with its characteristic dark rim composed of deposited coffee particles. (From Ref. 9 © 2000 The American Physical Society.)

Sommer and Rozlosnik,[12] who used extremely smooth, hydrophobic substrates with a total surface roughness smaller than 0.3 nm, suggest that ring formation is independent of contact line pinning, and that the forming ring of material is the cause of contact line pinning. Finally, Hu and Larson propose that coffee ring formation requires a pinned contact line, particles that adhere to the substrate, a high evaporation rate near the droplet's edge and the suppression of the Marangoni effect that results from the latent heat of evaporation.[13]

It can be readily seen that the phenomenon of coffee staining has generated a number of opinions as to the exact conditions that cause it. However, it is generally agreed that when a ring of material forms from an evaporating droplet of suspension, the particles that form the ring are carried there by a strong replenishing flow, which originates from the droplet's centre and is driven by evaporation of carrier solvent at the stationary contact line. The strength of the replenishing flow has been demonstrated by Magdassi et al.,[14] who found that in rings formed from the room temperature evaporation of aqueous droplets of silver nanoparticles that the electrical conductivity was 15% of bulk silver.

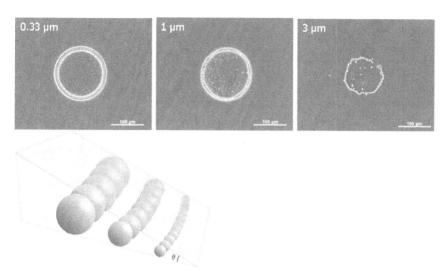

Fig. 3.   Phase contrast microscopy images of silica particle coffee stains on glass. The particle size is 0.33 μm, 1 μm and 3 μm from left to right. The coffee stains were formed from aqueous droplets, which were produced by inkjet printing. It can be seen that the dried droplet composed of 3 μm particles has the smallest diameter. The cartoon illustrates that smaller particles can deposit nearer to the contact line. (From Ref. 15, © 2008 The Royal Society of Chemistry.)

Coffee staining has also been investigated using an inkjet printer, which produced multiple aqueous droplets that had the same volume and were composed of uniformly sized silica particles.[15] This study showed that the size of a suspended particle influences the final dried morphology of a printed feature. If the contact angle formed by the droplet is less than 90°, smaller particles are located closer to the contact line than larger particles (Fig. 3). This deposition of particles can be explained by the wedge shape of the drying droplet's edge, which physically limits the movement of particles towards the droplet's periphery.

Controlling Ink Behaviour on Unstructured Substrates for Droplets and Arrays

A drying film of ink can be considered as behaving like a very large droplet. Just as a hemispherical droplet dries to leave a circular coffee

(a)   (b)   (c)

Fig. 4.   Three inkjet printed polystyrene films. The first film (a) was printed using a butyl acetate solution. The second film (b) was printed using a 1:9 methyl benzoate/ethyl acetate solution with a print head speed of 6.25 mm/s. Both solutions contained 2% polystyrene and 0.05% Disperse Red 1 by weight. The third film (c) was prepared from the same film as (b) but printed at 38 mm/s. (From Ref. 16, © 2004 The Royal Society of Chemistry.)

stain, a square film will dry to leave a dark frame[16] (Fig. 4a). Similarly, the approach to preventing coffee staining in droplets can be used with success in films. Studies performed with polymer solutions,[17] demonstrate that the use of a binary mixture of solvents can eliminate the formation of ring stains (Fig 4b), if one of the solvents has a much higher boiling point than the other.

The given explanation is that coffee staining is prevented since the solvent composition at the contact line moves towards an increasing percentage of the higher boiling point solvent than in the bulk. This shift causes a decrease in the evaporation rate at the contact line and establishes a surface tension gradient. A flow is induced from regions with low surface tension to regions with high surface tension when the Marangoni number, M is sufficiently large. ($M = \Delta\gamma L/\eta D$, where $\Delta\gamma$ is the difference in surface tension, L is the length scale involved, η is the viscosity, and D is the diffusion coefficient.) De Gans and Schubert showed that only small concentration gradients are necessary for Marangoni flow, which leads to homogeneity in the droplet and a reduction in the concentration gradient between the contact line and the bulk.[17]

Solvent ratios need to be optimised for the particular system under investigation.[18] Similarly, optimal printing conditions also need to be found. Figure 4c illustrates the effect of using an unsuitably high print head velocity. The ideal dot spacing, which is the distance between the centres of two droplets, also needs to be found. Print head velocity and dot spacing influence the drying time experienced by a deposited droplet, their role and influences are discussed in more detail in the next section.

The Behaviour of Lines of Droplets on Substrate

When a series of droplets is printed to form a line, their behaviour is similar to the single droplet and the film in that they can display coffee staining. That is, the bulk of the material is deposited at the edges of the line. Attempts to exploit this phenomenon have been made in the production of parallel, narrow lines using a copper hexanoate solution;[19] a decrease in coffee staining was observed when solutions of increased concentration were used. In another study, a simple formula for predicting track width using the equilibrium contact angle was tested using five substrates (ranging from glass, 5.9° to Teflon™, 58.7°) and an inkjet printed silver solution.[20] ($w^2 = (\pi d^3/6\Delta x)/(\theta/4\sin^2\theta) - (\cos\theta/4\sin\theta)$, where w = predicted line width, Δx = dot spacing and d = droplet diameter.)

When an inkjet print head dispenses droplets at a frequency, f over a substrate moving with a speed, U then the ejected droplets will land with a centre-to-centre separation $\Delta x = U/f$. If Δx is too large, the droplets will not overlap. However, if Δx is less than a droplet's diameter for contact angles less than 90°, or Δx is greater than twice the radius of curvature for contact angles greater than 90° then a line will form.[21] Gao and Sonin found that separate drops formed at lower frequencies, as expected, for fixed U. When frequency was increased bulges appeared in the line, which grew in amplitude as frequency increased further.

Schiaffino and Sonin investigated the formation and stability of lines built up from inkjet-printed molten wax.[22] They found that a molten bead forms with parallel contact lines, which freeze while the majority of the bead is still largely liquid. The still-molten material is stable when the contact angle is less than 90° but unstable above this, which is consistent with the theory of Davis,[23] who derived the conditions needed for stability for three cases: A) lines with a fixed equilibrium contact angle and mobile contact lines; B) lines whose contact angle depends on the contact line speed, but reduces to an equilibrium value at zero speed; and C) lines whose contact lines are arrested in a parallel state while the contact angle is free to change. He showed that cases A and B will always be unstable at some disturbance wavelengths, but that case C will be stable if $\theta < 90°$, which is representative of a liquid bead with strong contact angle hysteresis.

Case A was represented by water, which was printed onto polymethyl methacrylate[22] for a set frequency at various substrate speeds, all of which were low enough for the deposited droplets to overlap. Large, unconnected sessile drops were formed instead, with their size found to be dependent on substrate speed and deposition frequency. Duineveld observed the same phenomenon for aqueous droplets of PEDOT/PSS (poly(3,4-ethylenedioxythiophene) doped with polystyrene sulphonic acid) printed onto CF_4 treated glass ($\theta_a = 97°, \theta_r = 32°$).[24]

Duineveld's main concern was investigating liquids with zero receding angles, namely PEDOT/PSS on a substrate with $\theta_a = 66°$ or $\theta_a = 24°$. The PEDOT/PSS solution was inkjet printed so as to form lines. Many of these lines were unstable since they exhibited a series of regularly spaced liquid bulges that were connected by a ridge of liquid (such as can be seen in Fig. 5). The instabilities occurred when θ_i, which is the calculated initial angle formed by the deposited

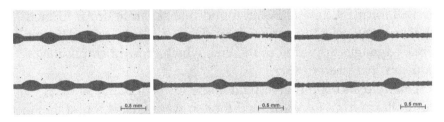

Fig. 5. Optical images of the bulges which appeared when printing TiO_2 ink on OTS-treated glass (contact angle $= 98.3°$) at 25°C, for $\Delta x = 0.05$ mm (left-hand side), 0.08 mm (centre) and 0.1 mm (right-hand side). (From Ref. 26, © 2007 The Royal Society of Chemistry.)

droplet, is larger than the advancing contact angle, θ_a of the line. This causes a bulge to form, which is connected to its neighbouring bulge by a liquid ridge. Fluid is pumped through the ridge due to a pressure difference that is generated by the forward moving liquid front at the head of the line, which causes the contact angle in the ridge to be smaller than the advancing contact angle.

A new bulge begins to form when θ_a is exceeded. The bulges were found to depend on both the substrate velocity and the applied liquid volume, with the distance between bulges decreasing for increasing substrate velocity and applied liquid volume. Instability in the line is also influenced by the transported flow rate through the ridge. Lines were found to be stable when the transported flow rate is much smaller than the applied flow rate.

When printing lines, one must consider the drying time experienced by each droplet, as recently demonstrated by Soltman and Subramanian,[25] who investigated the variation in morphology for printed lines as a function of varying Δx and substrate temperature. They found that bulges occurred when there was a short delay between droplet deposition and a small dot spacing (Fig. 6). Bulges also occurred for low temperatures and small dot spacings.

When low temperatures are used the ink droplets remain liquid for longer times, with successive deposited droplets adding to the

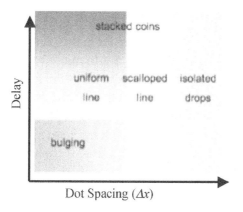

Fig. 6. Typical printed line behaviour of PEDOT:PSS ink at intermediate temperatures (approximately 30°C). (From Ref. 25, © 2007 The American Chemical Society.)

ink body. Similarly, a decrease in dot spacing can be thought of as an increase in the amount of ink per unit area. The extreme of going to higher temperatures and/or longer delays leads to each individual droplet drying before the next droplet is deposited, which also produces an unwanted morphology: stacked coins.

Finally, it has been shown that in some cases, the use of a lower critical solution temperature colloidal stabiliser can control ink behaviour on the substrate.[26] Above 37°C, the solubility of the stabiliser decreases causing a dramatic increase in viscosity. Lines printed using this approach did not display deviations at their starts and ends, and bulges in the line were prevented.

CONCLUSIONS

In conclusion, several factors need to be considered. When an ink droplet lands on a substrate it will spread to a greater extent for larger values of We. Afterwards, the droplet will lose mass due to the evaporation of its carrier solvent. Drying ink droplets tend to produce coffee stains, i.e. the majority of the material is deposited at the edge. This staining behaviour can also be seen for films. An effective approach for preventing coffee staining involves the use of a solvent mixture, one of the solvents has a much higher boiling point than the other; the exact ratio needs to be optimised for the system under investigation.

The substrate is not a passive partner in the printing process. Line width is determined by the contact angle made by the ink. Also the evaporation rate of droplets is affected by the substrate's thermal properties. Inkjet printed lines can exhibit coffee staining; they can also exhibit bulging or other unwanted morphologies. These phenomena can be corrected if one considers the evaporation profile of the drying droplets that make up the line. If the temperature is too high and/or the delay between each droplet is too long then each deposited droplet has dried before the landing of the next. However, if the temperature is too low, or the delay between droplet deposition is too short, or the dot-spacing is too narrow then the line will tend to exhibit regular bulges.

REFERENCES

1. Frohn A, Roth N. (2000) *Dynamics of Droplets*, Springer, Berlin.
2. van Dam DB, Le Clerc C. (2004) Experimental study of the impact of an ink-jet printed droplet on a solid substrate. *Phys Fluids* **16**: 3403–3414.
3. Pasandideh-Fard M, Qiao YM, Chandra S, Mostaghimi J. (1996) Capillary effects during droplet impact on a solid surface. *Phys Fluids* **8**: 650–659.
4. Schiaffino S, Sonin AA. (1997) Molten droplet deposition and solidification at low Weber number. *Phys Fluids* **9**: 3172–3187.
5. Park H, Carr WW, Zhu J, Morris JF. (2003) Single drop impaction on a solid surface. *AIChE Journal* **49**: 2461–2471.
6. Bourgès-Monnier C, Shanahan MER. (1995) Influence of evaporation on contact angle. *Langmuir* **11**: 2820–2829.
7. David S, Sefiane K, Tadrist L. (2007) Experimental investigation of the effect of thermal properties of the substrate in the wetting and evaporation of sessile drops. *Colloids Surf A Physicochem Eng Asp* **298**: 108–114.
8. Kavehpour P, Ovryn B, McKinley GH. (2002) Evaporatively-driven Marangoni instabilities of volatile liquid films spreading on thermally conductive substrates. *Colloids Surf* **206**: 409–423.
9. Deegan RD, Bakajin O, Dupont TF, Huber G, Nagel SR, Witten TA. (2002) Contact line deposits in an evaporating drop. *Phys Rev E* **62**: 756–765.
10. Deegan RD, Bakajin O, Dupont TF, Huber G, Nagel SR, Witten TA. (1997) Capillary flow as the cause of ring stains from dried liquid drops. *Nature* **389**: 827–829.
11. Deegan RD. (2000) Pattern formation in drying drops. *Phys Rev E* **61**: 475–485.
12. Sommer AP, Rozlosnik N. (2005) Formation of crystalline ring patterns on extremely hydrophobic supersmooth substrates: extension of ring formation paradigms. *Crystal Growth Des* **5**: 551–557.
13. Hu H, Larson RG. (2006) Marangoni effect reverses coffee-ring depositions. *J Phys Chem B* **110**: 7090–7094.
14. Magdassi S, Grouchko M, Toker D, Kamynshny A, Balberg I, Millo O. (2005) Ring stain effect at room temperature in silver nanoparticles yields high electrical conductivity. *Langmuir* **21**: 10264–10267.

15. Perelaer J, Smith PJ, Hendriks CE, van den Berg AMJ, Schubert US. (2008) The preferential deposition of silica micro-particles at the boundary of inkjet printed droplets. *Soft Matter* (in press).

16. Tekin E, de Gans B-J, Schubert US. (2004) Ink-jet printing of polymers — from single dots to thin film libraries. *J Mater Chem* **14**: 2627–2632.

17. de Gans B-J, Schubert US. (2004) Inkjet printing of well-defined polymer dots and arrays. *Langmuir* **20**: 7789–7793.

18. Tekin E, Smith PJ, Hoeppener S, van den Berg AMJ, Susha AS, Feldman J, Rogach AL, Schubert US. (2007) Inkjet printing of luminescent CdTe nanocrystal/polymer composites. *Adv Funct Mater* **17**: 23–28.

19. Cuk T, Troian SM, Hong CM, Wagner S. (2000) Using convective flow splitting for the direct printing of fine copper lines. *Appl Phys Lett* **77**: 2063–2065.

20. Smith PJ, Shin D-Y, Stringer JE, Derby B, Reis N. (2007) Direct ink-jet printing and low temperature conversion of conductive silver patterns. *J Mater Sci* **41**: 4153–4158.

21. Gao F, Sonin AA. (1994) Precise deposition of molten microdrops: the physics of digital microfabrication. *Proc R Soc Lond A* **444**: 533–554.

22. Schiaffino S, Sonin AA. (1997) Formation and stability of liquid and molten beads on a solid surface. *J Fluid Mech* **343**: 95–110.

23. Davis SH. (1980) Moving contact lines and rivulet instabilities. Part 1. The static rivulet. *J Fluid Mech* **98**: 225–242.

24. Duineveld PC. (2003) The stability of ink-jet printed lines of liquid with zero receding contact angle on a homogeneous substrate. *J Fluid Mech* **477**: 175–200.

25. Soltman D, Subramanian VS. (2008) Inkjet printed line morphologies and temperature control of the coffee ring effect. *Langmuir* in press.

26. van den Berg AMJ, de Laat AWM, Smith PJ, Perelaer J, Schubert US. (2007) Geometric control of inkjet printed features using a gelating polymer. *J Mater Chem* **17**: 677–683.

CHAPTER 5

Tailoring Substrates for Inkjet Printing

Moshe Frenkel

Digiflex, Israel

INTRODUCTION

Inkjet Inks & Print Quality

Print head technology determines inkjet inks' major characteristics, the latter being covered in the chapter: "Inkjet inks' Requirements". A major characteristic, substantially differing from standard printing inks, is the viscosity. While the viscosity of printing inks vary from several hundred up to several thousand cPs (centi Poise) — the viscosity of inks used for inkjet applications is in the range of a few cPs for the SOHO (Small Office and Home) printers and up to 20 cPs for commercial printers using industrial print heads.

A major implication of the ink's low viscosity is its high mobility on top of the substrate and the significant effect it has on print quality. On top of a non-porous substrate, ink might cause "clusters" in the case of under-wetting, high "dot gain" in the case of over-wetting, and "bleeding" at the interface of two adjacent colors.

The various effects of inkjet ink on top of a non-porous substrate are illustrated in Fig. 1.

A severe clustering effect on a non-porous material is demonstrated in Fig. 2.

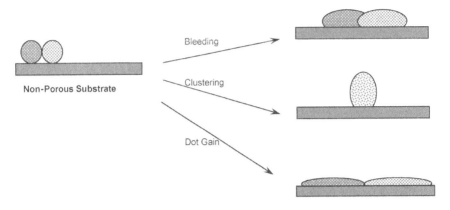

Fig. 1.   Inkjet ink behavior on top of a non-porous substrate.

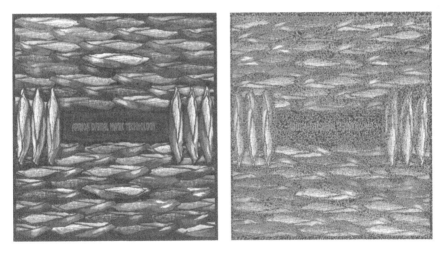

Fig. 2.   Clustering effect on print quality. Left: Original; Right: Inkjet print on non-porous substrate.

The picture on the left is the original, on the right is an inkjet print using an Epson home printer, on a non-porous, hydrophobic substrate.

Using Porous Substrates — here inkjet ink will penetrate into the pores of the substrate very quickly, causing low optical density at the surface and a potential "strike through" phenomenon when ink penetrates all the way to the other side of the substrate. Formation

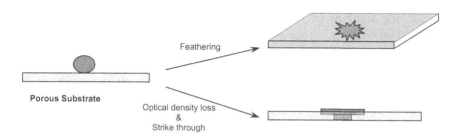

Fig. 3. Inkjet ink behavior on top of a porous substrate.

of a non-spherical dot, known as "feathering", is the result of ink absorption along the pores of the substrate. These phenomena are shown in Fig. 3.

This section might be well summarized, quoting Sandy Morrison[1]: "The inkjet paradox: Although it is clear that inkjet is the technology which has the widest range of applications and the greatest potential for future development, the paradox is that (as any user knows) it is not particularly good at printing on the most common substrate of all, plain paper. Toner-based systems are far more tolerant and capable of delivering clean images on colored or rough-surfaced papers. The reason is clear enough: Inkjet inks must have a low viscosity in order to jet successfully, and the spreading of the ink on a surface is dependent on how the water, oil, or solvent is absorbed and how rapidly the color is fixed to the surface. Thus, in addition to developing inks of higher quality, a great deal of effort has been put into developing substrates which are better suited to inkjet printing."

The various solutions offered to ensure high print-quality inkjet prints are reviewed below.

PRINT QUALITY SOLUTIONS

Matching Ink & Substrate

Some of the issues mentioned above, e.g., over-wetting, clustering, or bleeding can be eliminated by matching surface tensions of all ink colors (thus eliminating intercolor bleeding) and matching the surface tension of the ink to the surface energy of the substrate. This way no over-wetting (or, practically, high dot-gain) or clustering will be experienced, as well as eliminating bleeding phenomena.

However, an exact match of the surface tension of the inks and substrate means that a particular set of inks will be suitable for printing on narrow span of substrates — only the substrates matching the surface tension of the inks. Different substrates will require a different set of inks. Furthermore, even a full match of surface tensions (of both inks and surface) will not eliminate the issues which are related to porous substrates — including the most commonly used substrates — Paper! In order to enable high quality inkjet printing on paper, a whole set of new products was introduced to the market — "inkjet paper" or "inkjet substrate".

Inkjet Substrate

Inkjet substrates typically contain a special thin coating on top of paper or other substrates enabling high print quality. An inkjet substrate must demonstrate several unique properties to produce high quality images with inkjet inks. Once the ink droplet is deposited on the substrate, the ink must adhere without running or smudging and spread uniformly in all lateral directions to generate sharp edges. The substrate should also have adequate smoothness to promote high print densities. The paper should minimize bleeding and wicking tendencies, while enabling the quick absorption of ink onto the paper ("fast drying"), allow the dye to be well absorbed and fixed on top of the paper and be as thin a layer as possible to allow high optical density at the top of the substrate.[2]

The Principles

A substrate is coated with a thin, highly absorptive layer to absorb the ink droplets being ejected from the print head. The principle of the technology is presented in Fig. 4.

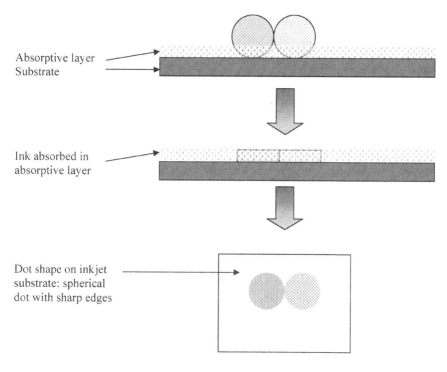

Fig. 4. Ink droplet absorption on "inkjet" coating.

The Technology

The main challenge of achieving all the abovementioned properties of the inkjet substrate is focused on the coating layer itself, both the composition and the formulation of the absorptive layer, as well as the coating technology. There are currently two such technologies offered for the coating types: microporous coatings and swellable coatings, both produced by coating paper with an impervious layer, then applying further coatings to produce the required properties.

a. Microporous coatings

The coating layer comprises small porous particles, demonstrating a very high absorptive capacity. The particles will absorb the ink droplet very efficiently and eliminate the undesired effects mentioned above. Inorganic particles as well as polymeric binders are used for this purpose. Most common inorganic particles include silica, alumina, calcium carbonate, and natural as well as synthetic clay minerals. It is the manufacturing process of these particles that will determine the particle size, shape, and porosity. Particles, both inorganic and organic, having a charged surface might be considered when used with ink containing ionic dyes. An opposite charge will help in fixing the dye to the particle surface. Using organic particles as the porous part in the ink receiving layer has been reviewed in the past.[3]

The major advantage of the microporous coatings is the speed with which they absorb inks as a result of their porosity. This instantaneous absorption provides, in effect, instant drying. Additional characteristics of the microporous layer are the high water fastness and its compatibility with both dye and pigment inks.

Glossy but highly absorbent microporous surfaces require thick coatings, which are typically composed of amorphous silica or alumina to provide the required absorbency. Due to the high surface area of the microporous layer, the ink dyes are distributed very thinly over a large surface area of the coating layer. This is the reason for the low gas fastness which this technology offers: even ozone levels as low as 0.1 ppm can produce a rapid loss of color, though performance has been greatly improved by recent developments in both dyes and coatings.

b. Swellable coatings

Swellable coatings technology consists of using special polymers and resin having a high capacity for swelling, on top of the paper (coating thickness varies at around 20 μm) to rapidly swell and absorb the ink and maintain high print quality. Typical polymers are gelatin, PVA, PVP, and cellulose derivatives which will absorb the ink droplet into the polymeric layer, thus eliminating its exposure to the atmosphere.

One of the reviews in this field, describing various polymers that are used as swellable coatings, was provided by Yuan et al.[4] The major advantage of this technology is the high gas fastness that it provides to the dye which is swelled into the polymer, eliminating its direct exposure to the atmosphere. The polymeric receiving layers are inherently glossy, while the porous particles cause a matte appearance of the substrate. In the past the only way to achieve high surface gloss was by coating it with a polymeric layer but recently, using the multi-layer structure of the receiving layer, high gloss is achieved when using porous particles as well. On the other hand, the major disadvantage of this type of coating is the relatively slow drying rate of ink which is governed by the swelling mechanism of the ink (via molecular diffusion) into the polymer, also known as "slow drying".

The characteristics of both technologies are demonstrated in Fig. 5 (the scheme taken from data provided by Ilford).

Some trends in the development of inkjet paper have been recently reviewed,[5] including microporous and swellable systems, which allow a high degree of ink absorbance on top of the paper and demonstrate high print quality of inkjet printing. A whole chapter in *Handbook of Imaging Materials* is dedicated to papers for inkjet printing[6] and describes the various technologies currently used.

Several studies were recently carried out to better understand the mechanisms which dominate the behavior of an ink droplet on top of plain paper or inkjet paper and to enable better design of advanced inkjet substrates. A detailed study about the spreading on and penetration into thin, permeable print media and its application to inkjet printing was recently published.[7] A detailed review of the interaction

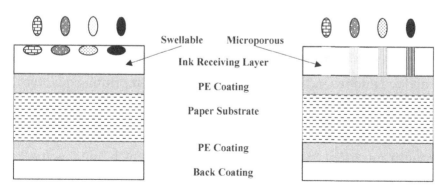

Fig. 5. Characteristics of "swellable" absorbing layer and "microporous" absorbing layer.

of both pure liquids and surfactant containing solutions with porous substrates is given for individual spreading and penetration and for the combined processes. A new model based on energy arguments is derived and compared to current hydrodynamic equations used to describe simultaneous spreading and penetration. The paper proposes a new model for the simultaneous spreading of surfactant solutions, which is based on energy arguments.

In "A Selected Overview of Absorption Criteria Derived from Recent Pore Network Modelling",[8] Gane claims that it is the absorption of fluids into porous media that controls the printability characteristics of coated papers with respect to ink setting rate and phenomena such as wicking and bleeding. It is the pore structure, size distribution and surface chemistry of the paper coating which govern the dynamic absorption of ink onto the paper. Gane reviews and discusses various models and theories which have been developed for the understanding of this phenomenon. He concludes that ink setting phenomena are complex and can be broken down into three main interactions: Firstly, the absorption of free fluid which can be described by a network model of pores and throats; secondly, the free fluid defined by the network wetting model and thirdly a chromatographic separation of different fluids making up a mixed fluid phase. Gane claims that it has been shown how all these factors can be modeled and measured by using very recently developed novel techniques. Applying these techniques can, in principle, provide for the systematic optimization of coating structures for the complete range of fluid-based printing technologies.

The use of specialty pigments in high-end inkjet paper coatings was thoroughly studied.[9] According to this study, inkjet papers can be divided into several categories according to their printing quality: The low- and middle-end papers are treated with different grades of surface sizing agents while the high-quality inkjet papers (high brilliance, round dot shape, high optical density, reduced strikethrough) are coated with a thin absorptive layer which contains specialty pigments and additives. This study examines how each of the coating components (e.g., silica pigments, precipitated calcium carbonate, polyvinyl alcohol) affects the coating properties.

Below is a short description of some individual inkjet paper technology examples which have been developed by the major players in the market.

a. Ilford[10]

Ilford uses nearly exclusively resin-coated true photobase for their wide format and desktop media. The ink-receiving layers (IRL) are based on proprietary technology and are applied in a multi-layer photo-like cascade or curtain coating process. Ilford developed ink receiving layers which are based on either polymers (e.g., gelatin; CMC, PVA) or microporous layers based on silica or lanthanum doped boehmite, or a combination of both.[11]

Ilford is calling the microporous layer "nanoporous" since the particle and pore size in the layer they produce is well below the micron level, with typical 20 nanometer particles while, practically, no particles are larger then 70 nanometers. The mineral oxides used in the porous products are surface treated in a proprietary process to balance physical properties such as brittleness and gloss with imaging properties such as color brilliance, layer transparency, and permanence.

A typical embodiment for the porous layer technology is described in several patents and patent applications, e.g., a US patent application in 2006.[12] This patent application describes a method for the preparation of silicon dioxide dispersions wherein the surface of the silicon dioxide is modified by treatment with the reaction products of a compound of trivalent aluminum with amino-organo-silane. The invention relates to recording sheets for inkjet printing having such a dispersion incorporated in the porous ink-receiving layer. Another US patent[13] describes the preparation of nanoporous alumina oxide or hydroxide which contains at least one element of the rare earth metal series with atomic numbers 57 to 71.

The Ilford polymer layer coating comprises a polyethylene coated paper support which is coated onto the front side of the said support, the ink receiving layer being a mixture of gelatin and rice starch.[14] Ilford products use multi-layer materials, containing polymer dry layers as well as microporous layers, a top coat which optimizes

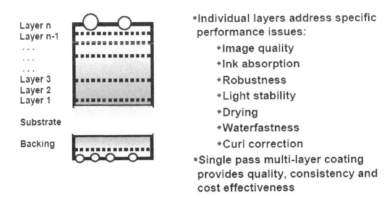

Layer n
Layer n-1
. . .
. . .
. . .
Layer 3
Layer 2
Layer 1
Substrate
Backing

▪Individual layers address specific
performance issues:
 ◆Image quality
 ◆Ink absorption
 ◆Robustness
 ◆Light stability
 ◆Drying
 ◆Waterfastness
 ◆Curl correction
▪Single pass multi-layer coating
provides quality, consistency and
cost effectiveness

Fig. 6.   Multi-layer ink absorbing structure model.

gloss, handling and friction as well as a backing layer which contains materials to prevent curling of the paper. A multi-layer design of an Ilford product approach is described in Fig. 6.

b. Kodak

A thorough review, published by Kodak,[15] describes one of their products — the Ultima Picture Paper which is intended to be used with inkjet inks. This is a general description which very well demonstrates the use of each of the building blocks we have been reviewing, in a commercial product. The basic structure (cross-section schematic) of the paper is described in Fig. 7.

The following is a brief description of each layer, from top to bottom.

1. Protective overcoat layer
    - A thin, highly ink-permeable polymeric layer.
    - Provides a non-tacky, abrasion-resistant surface after drying.
    - Contains a proprietary ceramic nanoparticle, a unique composition of matter that provides improved density and differential gloss.

2. Humectant management layer
    - Comprised of a swellable polymer and novel cationic resin.
    - Provides improved compatibility of the paper across a broader range of ink formulations.

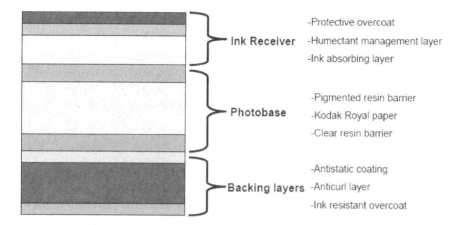

Fig. 7.   Basic structure of "inkjet paper".

- Serves as both a humectant "sponge" and a weak dye-fixing layer.

3. Solvent-absorbing base layer

- A thicker layer designed to rapidly absorb the primary ink solvent, water.
- Comprised of a swellable polymer and a novel cationic latex particle.
- Provides for additional, stronger dye fixing capability.
- Also contains microscopic beads to control the surface.
- A special "matting" agent is added.

The above three layers form what is collectively referred to as the "ink-receptive formulation". Critical to the success of this product is the COLORLAST technology embodied by the cationic polymer additives called "mordants" that are designed to bind and fix the dye molecules, along the ceramic nanoparticle in the protective overcoat. The exact choice, concentration, and location of the mordants are critical to achieving the best balance of image stability across the four main environmental factors: light, heat, humidity, and ozone. Equally important is incorporation of the proprietary ceramic nanoparticles present in the overcoat layer.

Taken together, this seventh generation ink-reception technology provides a more uniform gloss appearance between the printed and unprinted areas of the picture, greatly improved resistance to fading caused by light and pollutant gases such as ozone, and faster apparent drying times and improved smudge resistance of the final picture.

4. Pigmented resin barrier layer

- Extrusion coated polyolefin resin containing white pigments to provide an ultra smooth, ultra-white final substrate.
- The resin acts as a barrier to prevent the ink components from penetrating into the base paper, eliminating paper "cockle" and adverse interactions between the paper addenda and the imaging dye.

5. Base paper

- Produced from the finest wood pulps and state-of-the-art paper chemistry.
- Acid-free, non-yellowing.

6. Clear resin barrier layer

- Extrusion coated, clear polyolefin layer.
- Allows backprint to show through.
- Prevents water from absorbing into the back side of the base paper.

7. Antistat layer

- A very thin coating of proprietary antistatic material.
- Prevents charge build-up (static cling) during coating and printing.

8. Curl control layer

- Used to balance the natural curl imparted by the ink-receptive formulations.
- Comprised of a swellable polymer similar to that used in the ink-absorbing base layer.

9. Wet stacking layer

- Thinner, ink resistant layer coated over the curl control layer.
- Cross-linked polymer with matte particles to give it a textured surface.
- Prevents prints from sticking together when making multiple prints unattended.

c. FujiFilm

FujiFilm is also a very active player in the inkjet paper technology development. A short description of their technology is presented on their web site.[16] The nanoporous ink receiving layer has been refined to improve the smoothness of the surface while providing a luxurious gloss. This also leads to higher density of output images, producing the appearance of a deep sense of dimensionality.

The material for the ink receiving layer has been further refined and uniformly applied to the surface using thin-layer application technology. Image sharpness is improved by fixating the image on the surface of the paper. The layer has been made much clearer and more uniform to produce smoother granulation. Extremely high gloss is claimed to be due to a technology which produces very flat and smooth surface, appealing to the visual senses.

Technology Limitations

Inkjet substrate, "photopaper" or any of the brand names used for this type of technology, demonstrates a superb print quality when used with inkjet printers. However, there are two main limitations to this technology. The first is cost: the paper products of this type are extremely expensive — many users prefer to print their personal pictures in a traditional print shop, using photographic films, than printing the pictures at home using the inkjet paper. Due to the high cost of the product, the technology is not applied to wide format printing and very few companies are offering inkjet coating on top of polymeric substrates.

Heated Substrate

The Principles

Using inkjet special substrate coatings is, most probably, the most widely used method to enhance inkjet printing quality for the SOHO (Small Office & Home) markets. However, when dealing with commercial printing, e.g., wide format printers or non-porous substrates, top coats will neither guarantee the performance required by commercial customers, nor meet the cost goal required for most applications.

A method was developed to enable high print quality for those applications: "heated substrate". The substrate to be printed is heated during the printing process, to temperatures in the 40 to 70°C range. The mechanism which enables high print quality is driven by control of the ink viscosity and is explained in Fig. 8.

Two mechanisms will affect the ink drop once it hits the heated surface: decrease of viscosity due to the higher temperature of the ink and evaporation of the volatile ink components, causing higher solid concentration in the ink and a fast increase in ink viscosity. It is necessary to heat the substrate to a high enough temperature to ensure the viscosity-increase mechanism is dominant, to enable the overall viscosity-increase effect.

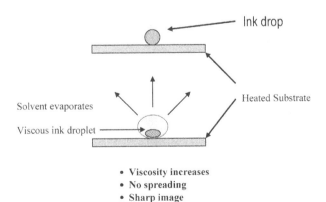

Fig. 8.   Ink droplet viscosity gain mechanism on a heated substrate.

Table 1. Temperature effect on ink dot spreading.

Substrate temperature (°C)	23	35	45	55
Dot Size (μm)	105	85	70	60

The increased viscosity of the ink will decrease the mobility of the ink droplet and will disable the mechanisms which result in poor print quality, ensuring the desired (high) print quality on top of the non-absorptive substrate.

The effect of temperature on dot size is demonstrated in Table 1: ink viscosity is 8cPs, drop volume is 22pl and substrate is grained anodized aluminum plate.

While the ink droplet spreads up to 105 μm when no heat is supplied to the substrate, spreading is very limited once the substrate is heated — the droplet viscosity increases and the spreading is limited accordingly. Most wide format printers, aimed to print on a variety of commercially available substrates, are using the "heated substrate" technology on flat-bed printers as well as on drum or roll printers.

Limitations of Heated Substrate Approach

The heated substrate approach is widely used where printing is done on non-absorbing media. Most wide format printers (also known as large format printers, ultra-wide format printers) have the capability of heating the substrate and this option is mainly used when printing on vinyl or similar substrates. However, when printing at high rates, substrate heating will become a major issue due to the high heating rate required by the substrate. Also, high substrate temperature will badly affect the print head by causing the ink to dry out at the nozzle and clog the printer.

"Bi-Component Ink" or "Reactive Ink"

The Principles

The "Inkjet Substrate" approach to enhancing print quality was characterized by a thin, highly absorptive layer, which physically absorbs the ink droplets, thus allowing uniform dot size, sharp and well defined dots edges.

Reactive ink is a different approach to achieving the same results via a chemical reaction. A thin layer of a reactant is introduced on top of the substrate to chemically react with a second reactant which is present in the ink. Once the ink droplet hits the substrate both reactants instantaneously react to practically "jell", "freeze" or immobilize the ink droplet or, in some approaches, the colorants present in the droplet. Once immobilized, the ink droplet will behave like an offset ink dot which does not tend to demonstrate any of the problems related to a low-viscosity ink, e.g., bleeding, strike-through, clustering or feathering.

The major difference between these two approaches ("freezing by absorption" *vis-à-vis* "freezing by chemical reaction") is the nature of the substrate coating. While ink absorption requires a thick enough layer to absorb the whole ink droplet into unique materials and pro-prietary coating methods, the chemical approach requires a very thin layer of the reactant to be introduced on top of the substrate (typi-cally < 1μm of dry layer). This layer might be introduced onto any substrate either before or as part of the printing process itself. The reactive-ink principle is demonstrated in Fig. 9.

Reactive inkjet inks are widely described in the patent liter-ature. Well over 100 patents have already been issued which claim different reactions between an ink component and substrate top-layer to enable high print quality, eliminating all undesired effects which are caused by the non-viscous inkjet ink. A breakdown of

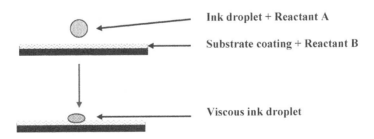

Fig. 9.   Reactive ink viscosity gain mechanism.

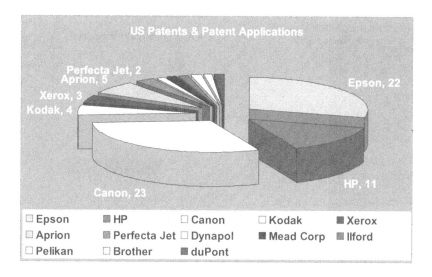

Fig. 10.   "Reactive ink" related patents, breakdown by companies.

about 75 US patents by the companies which applied for the patents is presented in Fig. 10.

HP has developed, in partnership with International Paper, what they call Paper Enhancement Technology.[17] This development, practically, uses a "bi-component" ink approach by introducing a "fixer" on top of the paper to react with the colorant in ink, fix it and keep it near the paper surface. This approach enables uniform optical density of the ink as well as well-defined ink dots, faster drying time and high optical densities, all done at low paper cost. HP Paper Enhancement Technology is described in Fig. 11.

Technology Approaches

The principle of this technology is a chemical reaction which causes the immobilization of the ink droplet or, at least, the immobilization of the colorants used in the ink. The various chemical approaches taken to achieve this goal are summarized in Table 2.

Breaking down the various patents to the approach taken, yields the scheme shown in Fig. 12.

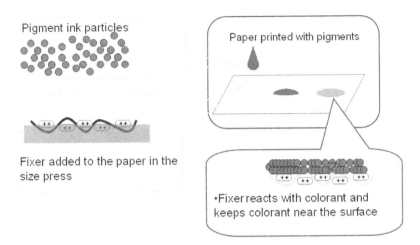

Pigment ink particles

Fixer added to the paper in the size press

Paper printed with pigments

•Fixer reacts with colorant and keeps colorant near the surface

Fig. 11.   HP Paper Enhancement Technology principles.

Table 2. Chemical approaches to achieving high printing quality using "bi-component" ink.

	Reactant in Ink	Reactant on Substrate	Reaction Product
I	Reactants in two adjacent colors	None	Bleed control
II	Anionic Dyes	Polyvalent metal ions; Proton	Precipitation of dye; Fixation
III	Resin in Ink	Polyvalent metal ions; Proton	Precipitation of resin & gelation of ink
IV	Dispersants in ink	Polyvalent metal ions	Breakage of dispersion; Fixation
V	Inorganic salt	Inorganic salt	Salt precipitation; Fixation

- Approach I: Reactants in two adjacent colors

This approach is aimed at eliminating intercolor bleeding phenomenon which takes place at the border of two adjacent colors when printed on a non-porous surface. Interaction between reactants which are present in two adjacent colors will cause either a viscosity increase in the ink ("jellation") or fixing of the colorants present in the droplet by converting them to insoluble compounds. The methods taken in this approach are summarized in Table 3.

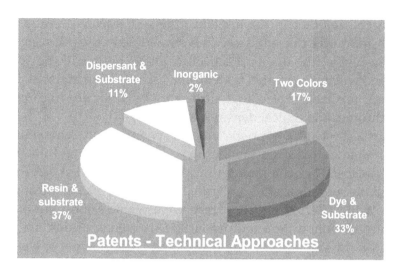

Fig. 12. "Reactive ink" related patents, breakdown by technical approach.

Table 3. Various chemistries used for 'reactants in two adjacent colors' approaches.

1st Ink Color	2nd Ink Color	References
Gel forming reagent	Gel initiating reagent	US Patent 6,777,462 (Xerox)[18]
Anionic dye	Ca^{++}; Cu^{++}; Cationic dye	US Patent 6,979,365 (Kao Corp)[19]
Resin soluble in ink	Resin precipitating cations	US Patent 6,281,267 (HP)[20]
Pigments stabilized cationically	Pigments stabilized anionically	US Patent 6,536891 (Epson)[21]
Pigments stabilized cationically	Anionic surfactants	US Patent 6,530,656 (Canon)[22]
Other approaches	Other approaches	US Patent 6,644,377 (Brothers Ind.)[23] US Patent 7,045,002 (DuPont)[24]

- Approach II: Anionic dye in ink and a "fixer" on substrate.

This approach will not only eliminate intercolor bleeding but might resolve all inkjet ink issues which have been reviewed above. Anionic dyes as the colorant in the ink and a second reactant, on top of the substrate, will cause the dye to be insoluble in the ink system. Being insoluble, the dye is practically "fixed" on top of the substrate and will not be carried with the ink carrier, either horizontally (meaning feathering, bleeding, clustering) or vertically (meaning strike-through and optical density loss).

Such a system comprises an ink composition, containing "effective amounts of an anionic dye, a polyamine, and an ink vehicle; a second pen containing an acidic fixer composition. The first pen and the second pen are configured to print the ink composition and the acidic fixer composition, respectively, onto a substrate such that the ink composition and the acidic fixer composition are in contact on the substrate."[25] Another patent "relates to a dye-based inkjet ink composition comprising: at least one anionic component, wherein, when the ink composition and a fixer fluid comprising at least one cationic component are mixed together to form a mixture, the mixture is an amorphous viscous fluid, the viscous fluid having a viscosity greater than the ink".[26]

- Approach III: Soluble resin in ink and polyvalent metal ions or a proton on top of the substrate.

Of the various approaches taken while using a bi-component ink system, this is the most common one as reflected in the patent literature (see Fig. 12). This system is based on the fact that several polymers are soluble in aqueous solution under certain conditions (e.g., pH value, presence or absence of metal ions) and will precipitate as soon as the chemical environment in the ink is changed. An acrylic polymer will be soluble in the ink at an elevated pH, hence it will instantaneously precipitate upon its reaction with a polyvalent cation which is present at the surface of the substrate. The result of this reaction will be a gelled ink droplet which will be practically immobilized. The schematic reaction mechanism is shown in Fig. 13.

This mechanism is used in various patents which differ in the materials used, method of applying the "reaction solution" on top of the substrate and additives used to improve the print performance.[27] One patent, issued by Epson Corporation, uses a resin in emulsion

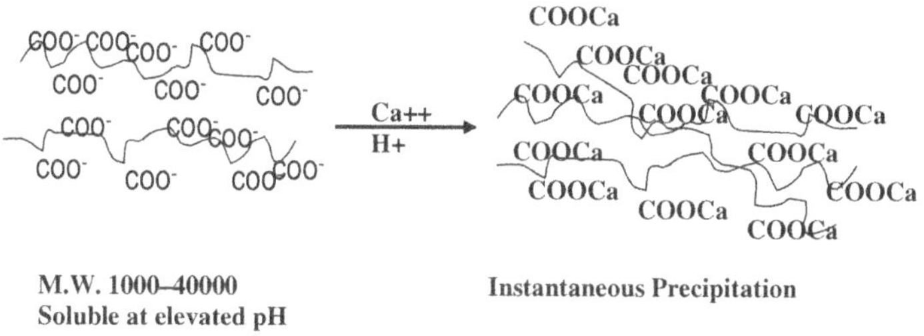

M.W. 1000–40000
Soluble at elevated pH

Instantaneous Precipitation

Fig. 13.   Acrylic polymer precipitation in aqueous solution.

which contains carboxyl groups at the surface and is precipitated by the polyvalent metal ions.[28] The use of a special swelling agent to allow the inorganic cations to partially penetrate into the non-porous polymeric substrate is described.[29,30] The same approach was taken to achieve high print quality when jetting water-based ink on top of offset plates.[31]

• Approach IV: Ionic dispersant in ink and polyvalent metal ion which precipitates the dispersant and breaks the dispersion.

Improving print quality via this mechanism deals with a dispersed ink system. The dispersion is stabilized by dispersants, which are surface active agents attached to the surface of the dispersed particles. The reactant which is present on top of the substrate ("reaction solution") is capable of breaking the state of dispersion of the ink (and, mainly the pigment dispersion or dissolution), making it practically immobilized.[32,33]

• Approach V: Reaction of two inorganic salts to precipitate an insoluble salt and cause fixation of ink droplet.

This approach was used to fix the ink droplet onto the substrate. One inorganic soluble salt is introduced to the ink while another soluble salt is coated on top of the substrate. The reaction product of the two salts is an insoluble salt (or salts), which, upon precipitation in the droplet, causes the immobility of the droplet to enable high print quality.[34]

Technology Limitations

The major drawback of this technology is that it requires the application of an extra layer on top of the substrate. This makes the printing process more complicated and will require either an extra coating step (when done off the inkjet printer) or a proper application tool in-line with the printer itself. Using bi-component ink, no matter which of the approaches is taken, it will still require drying of the ink vehicle — the water and solvent — before the ink is dry. Full drying is not required when using the absorptive layers.

SUMMARY

Print quality issues related to inkjet printing are common to all print heads technologies used in this field and to all ink compositions. These issues are derived from the non-viscous ink which is jetted onto all types of substrates. Various technologies have been developed to overcome the basic print quality issues and they have been reviewed in this chapter. Print quality improvement technologies include the absorptive layer coatings which are very widely used in the SOHO (Small Office and Home) market and are still under thorough development by the major companies. Heated substrate approach is used in the industrial wide-format and grand-format printers which use very large variety of substrates. The bi-component ink technology (or "reactive ink") was thoroughly reviewed, including the various approaches taken. At this point in time this approach is used only on one HP machine, but I believe it is one of the directions the inkjet technology will head in the next few years.

REFERENCES

1. Morrison S. (2005) Technology review: Digital print and inks. *Coatings & Inks* April issue Available at: www.specialchem4coatings.com/resources/articles/article.aspx.

2. Cawthorne JE, Joyce M., Fleming D. (2003) Use of chemically codified clay as a replacement for silica in matte coated inkjet papers. *J Coatings Technol* **75**: 937–975.

3. Nogucji Y, Satoh S, Fujii S, Hama R, Satoh M, Yamagishi M. (1999) Organic cationic, sub-micron particles for inkjet paper coatings. In Hanson F, *Recent Progress in Ink Jet Technologies*, pp. 370–374. Society for Imaging Science and Technology, Springfield, VA.

4. Yuan S, Sergeant S, Rundus J, Jones N, Nguyen K. (1997) The development of receiving coatings for inkjet imaging applications. *Proc IS&T NIP* **13**: 413.

5. Stephenson IR. (2004) Some trends in the development of inkjet paper and other digital printing technologies. *Surface Coatings International Part A* **87**(A3): 140.

6. Bugner D. (2002) Papers and films for ink-jet printing. In Diamond A & Weiss D (eds), *Handbook of Imaging Materials*, 2nd Edition, Chapter 15. Marcel Dekker, New York.

7. Daniel RC, Berg JC. (2006) Spreading on and penetration into thin, permeable print media: Application to inkjet printing. *Adv Colloid Interface Sci* **123–126**: 439.

8. Gane PAC. (2004) Absorption properties of coatings: A selected overview of absorption criteria derived from recent pore network modelling. *J Dispers Sci Technol* **25**(4): 389.

9. Hladnik A. (2004) Use of specialty pigments in high-end inkjet coatings. *J Dispers Sci Technol* **25**(4): 481.

10. The information was provided by Ilford with their permission.

11. Steiger R, Brugger PA. (1999) Photochemical studies on the light fastness of inkjet systems. In Hanson E (ed), *Recent Progress in Ink-Jet Technology*, pp. 321–334, Society for Imaging Science and Technology, Springfield.

12. Furholz U, Ruffieux V, Schar M. (2006) Recording sheet for inkjet printing, US patent application 2006 0078696.

13. Brugger P, Ketterer JK, Steiger R, Zbinden F. (2000) Recording sheets for inkjet printing, US Patent 6,159,419.

14. Fryberg M, Schuttel S, Tomimasu J. (2007) Recording sheet for inkjet printing, US Patent 7,235,284.

15. Bugner DE, Romano C, Campbell GA, Oakland MM, Kapusniak R, Aquino L, Maskasky K. (2004) The technology behind the new Kodak Ultima Picture Paper — beautiful inkjet prints that last for over 100 years. *Proc 13th International Symposium on Photofinishing Technology*, pp. 38–43. Reprinted with permission of IS&T: The Society for Imaging Science and Technology sole copyright owners of *IS&T 13th International Symposium on Photofinishing Technology Proceedings*.

16. FUJIFILM Global/Products/Inkjet Paper/Technology. http://www.fujifilm.com/products/inkjet_papers/technology/index.html.

17. Information was provided by HP and permission granted to publish it.

18. Smith TW, Colit RL, McGrane KM, Ly H. (2004) Ink compositions containing sodium tetraphenylboride, US Patent 6,777,462.

19. Tsuru I, Tsutsumi T. (2005) Ink set, US Patent 6,979,365.

20. Parazak DP. (2001) Ink-to-ink bleed and halo control using specific polymers in inkjet printing inks, US Patent 6,281,267.

21. Oyanagi T. (2003) Aqueous pigment-based ink set, US Patent 6,536,891.

22. Teraoka H, Takizawa Y, Yakada Y, Yakushigawa Y. (2003) Color inkjet recording ink set, inkjet recording methods, recording unit, ink-cartridge, inkjet recording apparatus and bleeding reduction method, US Patent 6,530,656.

23. Kawamura M, Kobayashi N, Ohira H, Higashiyama S, Furioka M. (2005) Ink set for inkjet recording, US Patent 6,866,377.

24. Bauer RD, Heransky CG, Hall WT. (2006) Interactive ink set for inkjet printing, US Patent 7,045,002.

25. Tsao YH. (2003) Enhancement of waterfastness using a polyamine/anionic dye mixture with an acidic fixer, US Patent 6,652,085.

26. Lee S, Byers G, Kabalnov A, Kowalski M, Chatterjee A, Prasad K, Schut D. (2006) Ink and underprinting fluid combinations with improved inkjet print image color and stability, US Patent 7,066,590.

27. Kazuaki W. (2000) Reaction solution for inkjet recording method using two liquids, US Patent 6,080,229.

28. Kubota K. (2002) Method for inkjet recording on non-absorbing recording medium, US Patent 6,426,375.

29. Nitzan B, Peled G, Schur N, Frenkel M. (2005) Pretreatment liquid for water-based ink printing applications, US Patent application 20050051051.

30. Nitzan B, Peled G, Schur N, Frenkel M. (2004) Surface treatment for printing applications using water-based ink, US Patent 6,833,008.

31. Nitzan B, Frenkel M. (2005) Pre-treatment liquid for use in preparation of an offset printing plate using direct inkjet CTP, US Patent 6,906,019.

32. Yatake M. (1998) Pigment ink composition capable of forming image having no significant bleeding or feathering, US Patent 5,746,818.

33. Koitabashi N, Tsuboi H, Fujimoto Y. (2002) Ink-jet printing method, US Patent 6,494,569.

34. Frenkel M. (2006) Ink composition and inkjet printing method using the same for improving inkjet print quality, US Patent Application 20060028521.

PART II

FORMULATION AND MATERIALS
FOR INKJET INKS

CHAPTER 6

Pigments for Inkjet Applications

Alex Shakhnovich and James Belmont
Cabot Corporation

INTRODUCTION

The inkjet (IJ) method of non-impact printing has been used commercially since the 1980s. Originally the inks contained colorants that were soluble in the ink vehicle, i.e., dyes. Later on, however, deficiencies of the dyes, such as mediocre image permanence (light- and weatherfastness, ozone resistance) and poor waterfastness became obvious. Industry requirements for better colorants for outdoor or photo printing were becoming more and more demanding. It is well known that in applications such as paint, plastics and coatings, the use of pigments instead of dyes increases the performance characteristics of the color package dramatically. The same turned out to be true for IJ. The use of carbon black instead of weaker black azo dyes totally revolutionized the IJ industry, bringing text printing close to laser printer quality regarding speed, edge acquity and durability. In fact, deficiencies in black were the main driving force for considering pigmented inks. By the late 90s all major players were using black pigment because of carbon black's desirable image permanence, high color strength and flat spectral absorption. Epson and Kodak now broadly use color pigmented inks in their IJ products, while HP, Canon and Lexmark still use color dyes and limit pigment use to high end photo and office applications.

Pigments by definition are not soluble in the media of application, so the waterfastness of pigment-based prints is generally much better than that of dye-based prints. As expected, the light- and ozone fastness of pigmented prints also turned out to be much improved compared to dyes. However, a number of significantly improved IJ dyes with high light- and ozone fastness have been discovered and developed recently.

In order for the pigments to be formulated into inkjet inks they have to be dispersed to small particle sizes (roughly between 50 and 200 nm, depending on the application) and this dispersion needs to be made colloidally stable. The colloidal stability can be achieved either by using pigment surface modification to form an adequate surface charge, or by adsorption of certain compounds on the surface of pigment particles. These compounds may be low-molecular weight pigment derivatives or they may be polymeric dispersants, stabilizing the particles by steric repulsion or by charge repulsion. The details of these methods will be described later in this chapter.

These improved properties often come at a cost. Replacing a colorant solution by a colorant dispersion raises several issues.

1. The colloidal stability of the dispersion is a critical property, as the inkjet ink should not settle for months or longer, depending on the application. Although all pigmented inks will eventually settle, their resistance to settling depends on their size, density, surface chemistry and the dispersants used. As the settling rate is proportional to the square of the diameter and to the density difference vs. the solvent, large particles are problematical. Reducing the particle size of pigments below 100 nm is usually technically challenging and costly. Some pigments such as Pigment Green 36 have high densities due to their metal and/or halogen

content and are extremely hard to maintain in a colloidally stable state.

2. Second, ejecting particulates through the microjets (may be 10–30 microns in diameter) at high velocity puts many restrictions on the ink properties. The viscosity and surface tension of the ink should be carefully controlled, which may be difficult in a system that contains both dispersants and nanoparticulates at high loading. Of utmost importance is the large particle content (usually expressed as number of particles larger than 0.5 or 1 micron per 1 ml of ink). Large particles may plug the jets and channels and cause irreversible damage to the print head.

3. In the case of thermal IJ print heads, where the surface of the firing resistor is briefly heated to a very high temperature, no deposition on the resistor should take place. Such a deposit is thermally insulating and usually damages the resistor irreversibly. Preventing this phenomenon adds additional requirements to the ink purity and composition.

4. Depositing pigment particles on the paper surface creates many problems related to the penetration, adhesion and cohesion of these particles, and sometimes results in poor highligter smear resistance, gloss inconsistency of photo prints and the like.

5. Inks with high solids contents are required for some lightfast pigments that have low color strength compared to dyes. This problem, however, seems to be limited mainly to quinacridones.

Most of the named problems are technical, rather than fundamental in nature. They can be solved or alleviated by carefully choosing the surface treatment, by ink formulation work and/or by selecting appropriate printing algorithms. Pigments are continuing to replace dyes for many non-impact printing applications, particularly IJ. IJ grades of pigments, IJ pigment dispersions or both are now commercially available from multiple global manufacturers, such as Cabot Corporation, Clariant, Ciba, DuPont, Dainippon Ink, Daicolor-Pope, Toyo Ink, Dainichiseika, Degussa, Sensient, Kao, Mikuni and others.

CARBON BLACK

Carbon black is produced by the partial combustion or thermal decomposition of hydrocarbons. Several methods are used, including the furnace black, thermal black, lamp black and acetylene black processes.[1-3] The furnace black process is the most common. In this process, natural gas (or another fuel) is burned to form a hot gas stream that is directed into a tunnel. An aromatic oil is sprayed in and the black forms as the gas moves down the tunnel. The reaction is quenched with the addition of water, and the product is collected as a low density powder (fluffy black) or is further processed into millimeter sized pellets.

Morphology

Carbon black aggregates consist of primary particles that are fused together to form fractal-like structures. The aggregates associate with each other because of van der Waals forces to form agglomerates and can be formed into millimeter sized pellets during manufacturing. The primary particles are typically 10–75 nm in diameter and the aggregates are typically 50–300 nm in diameter. The surface area of carbon blacks is largely determined by the size of the primary particles and is usually in the range of 30–200 m^2/g, but there are some outside these ranges with some exceeding 500 m^2/g. The aggregates have significant internal void space because of their fractal nature or "structure". The dibutyl phthalate absorption (DBP) number or oil absorption number (OAN) is used to characterize commercial carbon blacks and is a measure of the combination of the internal void volume and the volume between closely packed aggregates. These values normally range from 40–130 mL/100g of carbon black.

The primary particles of the carbon black are made up of amorphous carbon and small graphitic crystallites roughly 15–25A in size. TEM work has shown that the crystallites near the surface are parallel to it and the ones in the center are randomly arranged. There are different models of the actual surface.[4] In one, there are graphitic basal planes separated by regions of amorphous carbon. In another, the edges of graphitic crystallites are exposed at the surface with the basal planes buried inside the particle.

The morphological characteristics of carbon black affect its ease of use in a variety of applications and the properties of materials containing carbon black. For example, carbon blacks with low surface areas and high DBP values are more easily dispersed into inks and other media than blacks with high surface areas and low DBP values. The pigmentation properties also depend on morphology, with high surface area materials giving the greatest jetness (blackness), provided that the blacks are well dispersed. The undertone color depends on the primary particle size and the application.

Surface Chemistry of Carbon Black

Extensive efforts have been made to characterize the surface chemistry of carbon blacks.[1,5] Although carbon blacks are nearly all carbon, impurities of oxygen, sulfur, nitrogen and small amounts of other elements are present. Most of the work has centered around the identification and quantification of oxygen containing

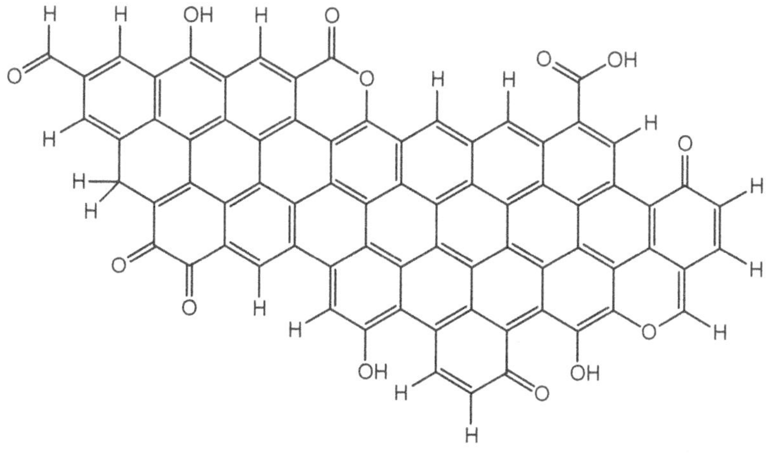

Fig. 1.   Functional groups on the surface of carbon black.

functional groups. Some approaches are based on titrations, while others use pyrolysis/MS,[6] FTIR,[7] Raman,[8] ESCA[9] or combinations of wet chemical and spectroscopic methods.[10, 11] Through this work, a number of functional groups have been identified as being present on the surface of carbon black including carboxylic acids, ketones, phenols, anhydrides, aldehydes, quinones, lactones, lactols and pyrones.[5, 12] Carbon blacks have a mixture of these groups on their surfaces (Fig. 1). The quantification of the groups remains challenging, though.

ORGANIC PIGMENTS

It is well known that a robust composite color match can be built from a white, a black and from two colorants (A and B) which have hue angles correspondingly smaller and larger than the required colors. The closer the hues of A and B are to the required color, the better the match will be. This is why the industrial color palettes for paints and plastics applications usually consist of 12–20 colorants, more or less evenly spaced by hue angle. As there is limited number of jets available, such a large palette is not practical for most of IJ applications.

The most widely used pigment set for IJ consists of a cyan (C), a magenta (M) and a yellow (Y) pigment. Together with a black pigment (K) that makes a 4-color set, commonly designated as CMYK. All composite colors are built by superposition of these four primaries (paper serves as a scattering white). That means that color reproduction may not be adequate for some demanding applications, such as photo prints. Therefore, some photo printers may have additional color cartridges such as green, orange, red, blue or, otherwise, two cartridges with the same colorants, but at different dilutions (light magenta, light cyan or grey). That approach may increase the number of cartridges to 8–12, which may add to the cost of the system considerably. Some typical a*-b* curves for a CMY inkjet pigment triade (PB15 – PR122 – PY74) are shown on Fig. 2 as a function of print density.

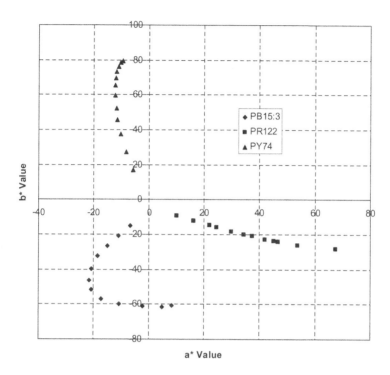

Fig. 2. Typical color values for 12-step monochromatic ramps printed on Epson Premium Glossy Photo Paper.

The 3-color CMY set is well defined; for a cyan color almost all ink manufacturers are using green shade copper phthalocyanine — Pigment Blue 15:3 or 15:4. This pigment is coloristically very strong, extremely lightfast and weatherable. Its hue angle is generally between 240° and 280°. The only fundamental problem with PB15 is its strong reflection in the red area,[13] which shows itself as bronzing (the reddish shade luster of a print when viewed at an angle). Although there are many inventions targeted to mitigate bronzing, it is a fundamental problem and cannot be eliminated completely.

Quinacridone pigments are the main class of colorants used as IJ magentas. Their permanence and clean color are responsible for the choice. Quinacridones, however, are relatively weak colorants. Their pink color is detemined not by molecular structure, but rather by association of molecules in the crystal and by intramolecular electron transfer. Therefore, the pigment loading in magenta ink is usually higher than in cyan or yellow inks.

PR122 R = Me
PV19   R = H
PR202 R = Cl

Many ink manufacturers use bluish red Pigment Red 122 (dimethylquinacridone). Pigment Red 122 has only one crystalline modification and its hue angle is generally between 315° and 340°. A recent trend, however is the replacement of PR122 with red shade (RS) PV19 (γ-modification of quinacridone). The reason for this is the difference in hue angles between these pigments — RS PV19 has a hue angle of between 355° (−5°) and 10°. Because RS PV19 is much yellower than PR122, it makes much cleaner composite reds and oranges. Another pigment that may be used in this group is PR202 (2,9-dichloroquinacridone). It usually looks stronger and yellower than PR122. Several attempts had been made to use red Naphtol AS pigments as magentas. However their lightfastness is so inferior to quinacridones, that this approach is now practically abandoned. The widest variety of pigments used in IJ is certainly in the area of yellow. That can be explained by the fact that as of today, there is no single yellow pigment which would meet all of the stringent IJ requirements. By far the most widely used IJ yellow pigment is

Pigment Yellow 74. It is a very strong and clean yellow colorant, widely available from many manufacturers. Its hue angle is slightly over 90° at full strength, making it a good choice for a neutral yellow. PY74 has two major drawbacks, though.

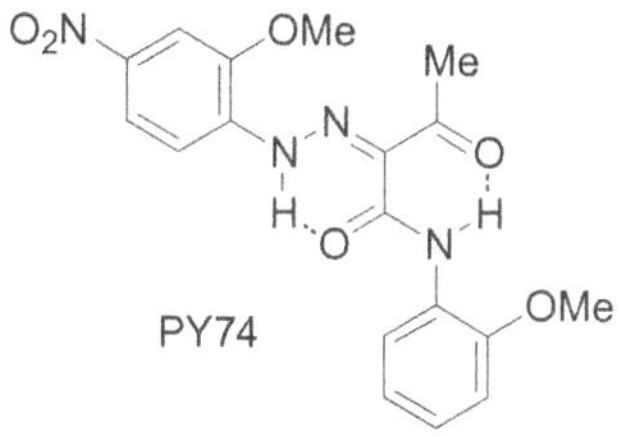

PY74

One of them is its poor lightfastness, which is especially noticeable when the pigment is dispersed to particle sizes below 100 nm. Another one is the slight solubility of this pigment in some components of ink vehicles, which makes formulation work more difficult and sometimes compromises the long-term colloidal stability of the inks.

There are several other azo pigments, which can be used to improve the properties of yellow ink dispersions.

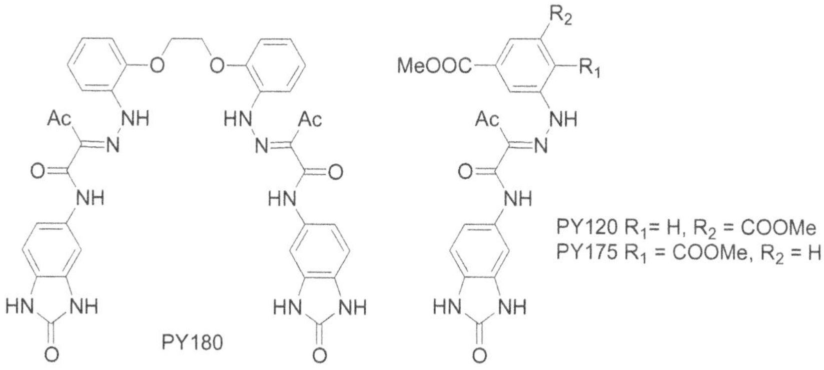

PY180

PY120 R$_1$= H, R$_2$ = COOMe
PY175 R$_1$ = COOMe, R$_2$ = H

Many attempts had been made to use benzimidazolone pigments, such as PY180, PY120 and PY175. They have much improved solvent resistance and higher lightfastness than PY74. Colorwise, PY180 is very close to PY74 and is almost as strong. However, the colloidal stability of these three pigments in dispersions is much more difficult to achieve because of the strong self-assembling action of the surface benzimidazolone groups. Inks based on benzimidazolone yellows are not commercially available. Another choice of a yellow azo-pigment is represented by dimeric PY155.

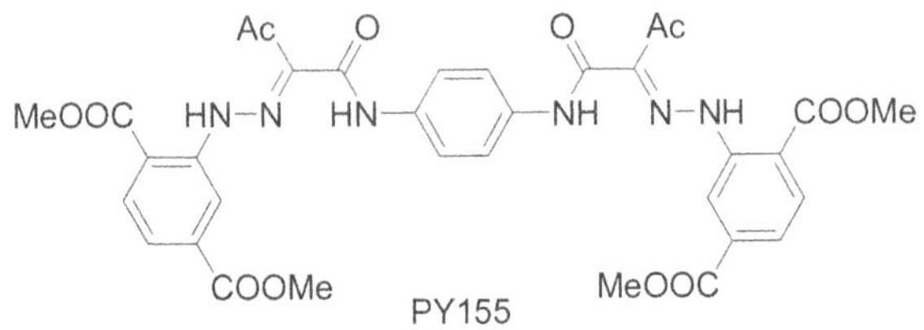

PY155

Being a little greener and weaker than PY74, this pigment has an improved lightfastness and is used in some commercial inks. Obviously enough, strong and inexpensive diarylide yellows such as PY12 are rarely mentioned in the IJ literature due to potential benzidine release,

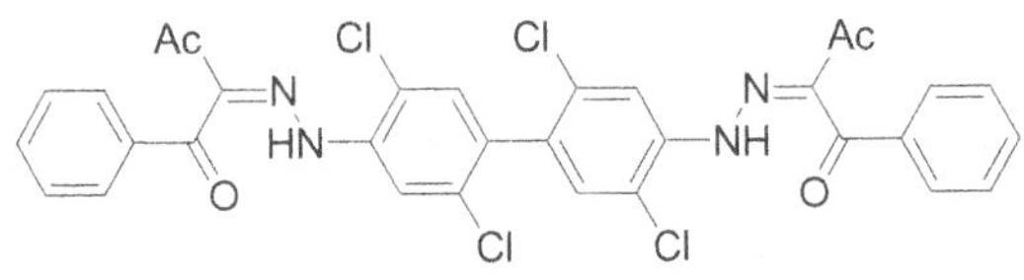

PY12

especially in thermal IJ systems. The most lightfast pigment in the azo yellow group is disazocondensation yellow PY128.

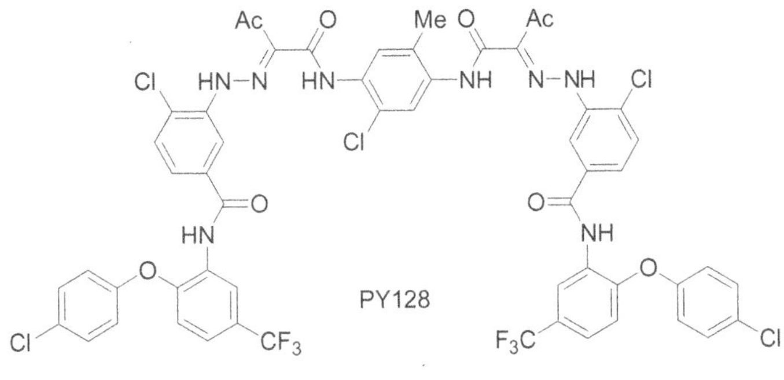

PY128

This pigment is weaker (per weight unit) than other azo pigments due to the presence of heavy side groups. It has limited use in some commercial lightfast inks. There are also some references to several heterocyclic yellow pigments, such as PY138, PY139 or PY185, as well as PY150 (metallized). One group of yellow heterocyclic pigments deserves special mention. These are quinolono-quinolones (QQ), mentioned in several Japanese and US patent documents.[14, 15] These pigments have been known since the 1930s, and commercially viable preparation routes have been disclosed in 1967–1968 and later improved in 2002–2007.[16]

PY220

PY218 R = F
PY221 R = Cl

Structurally, QQs are very similar to quinacridones. Their shades range from reddish yellow (2-Fluoro-QQ or PY220) to neutral yellow (3-Fluoro-QQ, or PY218 as well as 3-Chloro-QQ, or PY221). These pigments have outstanding lightfastness and solvent resistance, which are on par with phthalocyanines and quinacridones. Using them as a yellow component of a CMY triade allows the creation of an ink set for outdoor applications.[17]

DISPERSION METHODS FOR INKJET PIGMENTS (NANOSIZED DISPERSIONS)

Earlier in this chapter we outlined several strict requirements for IJ pigment dispersions. First and foremost, these are colloidal stability and particle size distribution. Generally the particle size has to be reduced to 80–150 nm, which is often challenging.

Particle size reduction may be achieved with different kinds of milling equipment such as media mills.[18] Alternatively, flow impingement or ultrasound methods can be used for dispersion. Some of the pigments can also be dissolved in certain solvents and reprecipitated as extremely fine particles.[19] All of these processes generate a considerable amount of fresh surface. If unstabilized, these surfaces collapse back together and particle size reduction may not happen. To prevent this collapse, different methods of stabilization of these surfaces are employed. The easiest way is to use commercially available pigment dispersants, which are widely available. These are either polyelectrolytes with affinity to the pigment surface, or polymers with bulky groups providing steric stabilization. Generally they are non-selective, so they work better with some pigments than with others. Their major advantages are their commercial availability and relatively low price. It is possible to "encapsulate" the pigment with polymer by changing solvent environment and/or pH to "crash" the polymeric dispersant onto the surface of pigment particles. In this case the polymeric dispersant is associated with the pigment by means of physical adsorption, so this process is reversible and the dispersion stability is extremely sensitive to components of inks. If these components are good solvents for the encapsulating polymer, the dispersion stability may be compromised, especially in the long-term. In general, the use of conventional dispersants makes the job of ink formulation difficult. Many examples of inkjet dispersions obtained by encapsulation are described in patents and applications by DIC. The polymeric dispersant used here is a co-polymer of styrene-acrylic acid-methacrylic acid with an acid number around 150. According to the inventors, this material is equally well suited for dispersing carbon black and color pigments.[20]

A custom polymeric dispersant is an improvement — the polymer molecule is tailored to the pigment and to the application. There are dozens of patents from companies such as Canon, Epson, KAO, DIC, DuPont and others describing the synthesis and use of custom pigment dispersants for IJ. Some of them are very complex, involving such techniques as grafting and block co-polymerization. For example, some DuPont patents describe a dispersant based on a diblock co-polymer A-B or triblock co-polymer A-B-A where block B has an affinity for the pigment, whereas block A is responsible for colloidal stabilization.[21]

A further improvement in the polymer dispersant area is possible with the appropriate design of a polymer-attached anchoring group, which has specific affinity for pigment surface. In several examples, chromophores such as AmDMQA were synthesized and attached to a styrene-maleic acid or styrene-acrylic polymer.

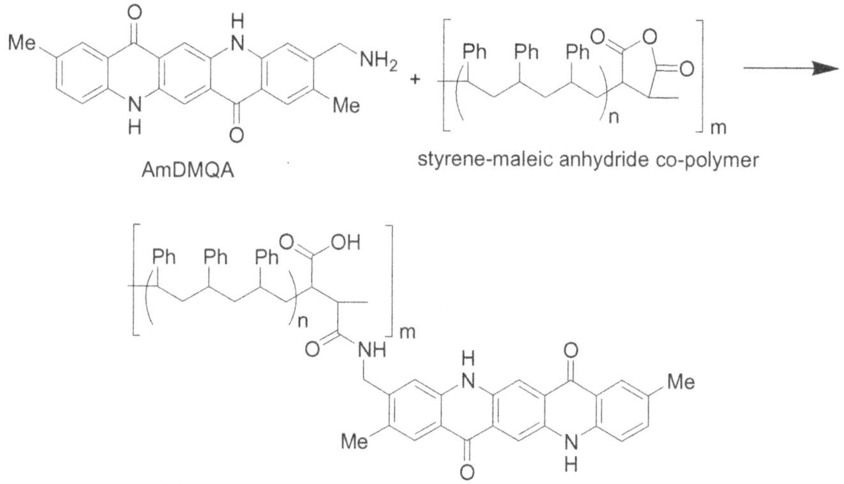

AmDMQA

styrene-maleic anhydride co-polymer

Dispersant with anchoring group

This custom dispersant is very effective in dispersing various quinacridones. In this case, the cooperative action of multiple anchoring groups within one polymer chain is beneficial for dispersion stability.[22]

Another method for colloidal stabilization makes use of low molecular weight pigment derivatives. Pigment derivatives are molecules which have structures that are similar to the corresponding pigments but contain bulky or ionizable groups. The rationale for pigment derivative use is the following: due to the similarity to the corresponding pigment structures, they adsorb well on the surface and therefore control the crystal growth. They are also called crystal growth inhibitors. Some examples are given below:

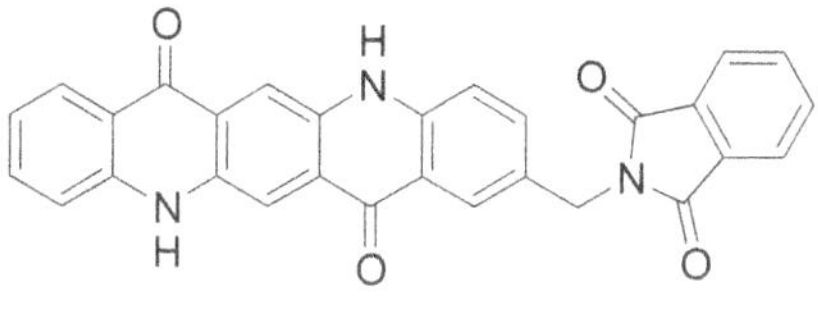

PMQ

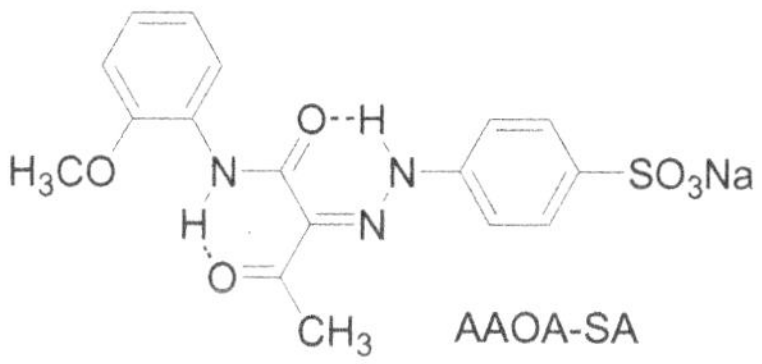

AAOA-SA

The use of PMQ for dispersing quinacridones is quite common and well documented. AAOA-SA is one of the many similar compounds that can be prepared from acetoacetanilides and aromatic amines which contain ionic or ionizable groups. Some of them are good dispersants for PY74 and related azo pigments. Several stable IJ dispersions have been generated; they gave images with good color saturation. A dispersant similar to AAOA-SA can be either generated in the presence of a presynthetized azo pigment to form an inkjet dispersion or, conversely, both the azo pigment and dispersant can be formed simultaneously in a mixed synthesis of PY74 or another azo pigment providing a self-dispersible IJ grade.[23] The dispersion properties can be fine-tuned by optimizing the diazonium salt structures, ratios and other conditions.

SURFACE MODIFICATION

The most robust method for changing the surface properties of pigments is to attach functional groups to the surface directly or through linking groups. The modifications can be chosen to improve the dispersion stability of IJ inks containing the pigments as well as to improve the printing properties of the inks.

Carbon Black

The oldest method for the modification of carbon black surface chemistry is oxidation.[5] Common oxidants include air, hydrogen peroxide, hypochlorites, nitric acid, nitrogen dioxide, ozone and persulfates. Each reagent produces a mixture of oxygen functional groups on the surface, with the distribution depending on the oxidant. Materials that disperse in water can be produced with sufficient oxidation, and hypochlorites[24] and persulfates[25] have been used to make water dispersible carbon blacks for inkjet inks.

Attached Organic Groups

The surface can also be modified via a reaction with organic molecules or intermediates to introduce specific functional groups.

One approach is to use the native oxygen groups as a reagent and site of attachment.[26] Usually, only one kind of group can be used for the reaction. As a result, only part of the surface is available for reaction, even when oxidized blacks are used. Another approach is to use the most abundant part of the surface, carbon, for the reaction. Several methods of this type have been explored including those based on diazonium chemistry,[27] azo chemistry,[28] peroxide chemistry,[28] sulfonation,[29] cycloaddition chemistry[30] and other methods.[31,32]

The reaction of carbon[27,33,34] with the diazonium salts is the most widely used method for surface modification of carbon blacks for inkjet inks.[35] The organic groups are attached directly to the carbon surface, and the surface coverage is not limited by the availability of specific oxygenated groups. The diazonium salt enables the introduction of a variety of functional groups onto the surface. A phenyl group is typically used as a spacer between the carbon surface and the functional group, but other spacers are possible. Simple functional groups such as sulfonate or carboxylate can be used to make treated carbon blacks with excellent dispersion stability and low surface tension in water and IJ inks.

Due to the flexibility of the chemistry, other properties can be controlled by adjusting the number, position or type of groups on the phenyl ring. For example, the optical density of prints made from inks containing the modified pigments can be improved when calcium binding groups are attached to the surface.[36] The groups may complex with calcium in paper under certain conditions, destabilizing the pigment and giving darker prints. One possible calcium binding group is the bisphosphonate group:

More complex groups can be attached to the carbon surface via secondary reactions with groups already attached through diazonium chemistry.[37] Polymers can be bonded to the pigment surface either via a direct diazonium reaction[38] or by a secondary reaction. The attached polymers can give improved durability to printed inks.

Organic Pigments

There are significant differences between organic pigments and carbon black that affect the ability to bond to their surfaces. First, organic pigments are molecular crystals. The molecules of the organic pigments are bound together with relatively weak bonds compared to the covalent bonds in carbon black particles. Consequently, the individual pigments have specific, well defined compositions. Further, most organic pigments are not good electron donors, as is carbon black.

There are several methods for modifying the surface of organic pigments. Some are effective for a limited range of pigments, while others are more general. An example of the former is the reaction of Pigment Green 36 with thiolates to form sulfide links to the surface.[39]

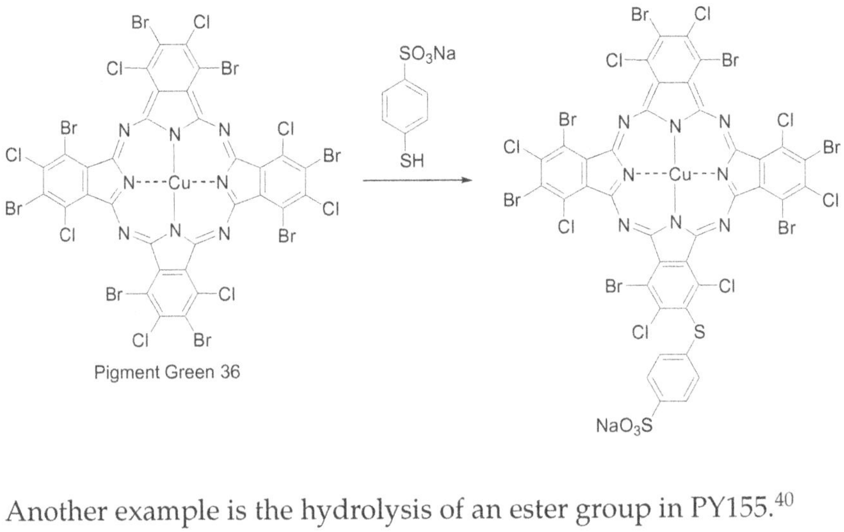

Pigment Green 36

Another example is the hydrolysis of an ester group in PY155.[40]

Modified PY155

Examples of more general reactions are sulfonation,[41]

chlorosulfonation or halomethylation followed by amidation,[42, 43]

and aralkylation using formaldehyde.[44]

Pigment modification with diazonium salts is also possible. Cabot successfully expanded the diazonium treatment method of carbon black, described above, to several color pigments.[45]

In an effort to elucidate the mechanism of diazonium treatment of the color pigments, the quinacridone PV19 was reacted with the carboxyphenyldiazonium ion. Nitrogen was evolved. Extraction of the product with dimethylformamide yielded seven red dyes with UV-VIS spectra very close to that of the parent PV19, which means that no change of chromophore had taken place. Three of the dyes were positively identified by MS as product of mono- and bis-arylation of PV19. Two of the structures are shown below and are examples of the expected structures.

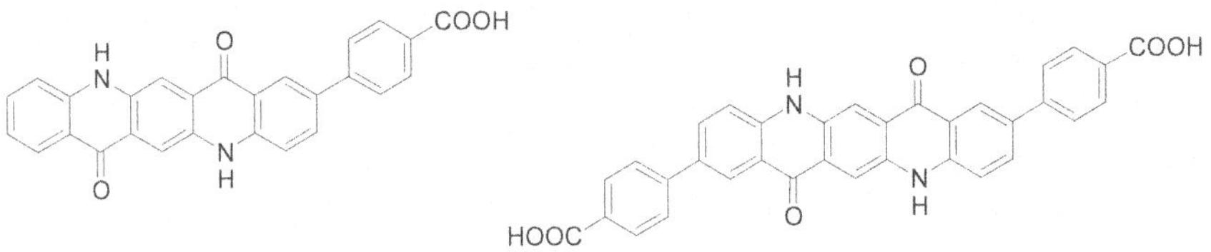

In the case of the reaction of the sulfophenyldiazonium ion with the quinacridone PR122, many products were again found. Among them was the azo dye shown below, indicating that some azo coupling can also occur to some extent.

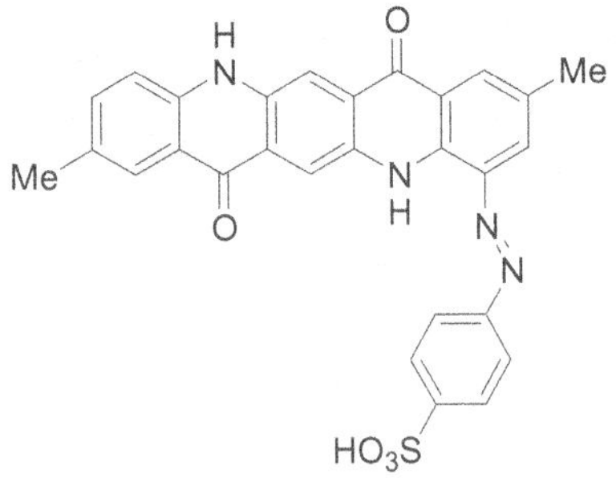

It should be noted that *soluble* acridone and quinacridone derivatives, such as ASA, QSA or DMQSA do not react with diazonium salts. This important finding indicates that the reactivity of the

PR122/PV19 surface depends on the solid state nature of the pigments.[46]

ASA

R = H; QSA
R = Me; DMQSA

Dispersions of Pigment Yellow 74 (azo pigment), Pigment Red 122 or Pigment Violet 19 (quinacridones) and Pigment Blue 15:3/4 (phthalocyanines), available under the trade name of Cab-O-Jet®, are now used as sources of CMY pigments by the inkjet industry.

REFERENCES

1. Dannenberg EM, Paquin L, Gwinnell. (1992) Carbon Black. In Kroschwitz JI, Howe-Grant M (eds.), *Kirk Othmer Encyclopedia of Chemical Technology*, Vol. 4, pp. 1037–1074. John Wiley & Sons, New York.

2. Donnet J-B, Bansal RC, Wang M-J. (1993) *Carbon Black*. Marcel Dekker Inc., New York.

3. Wampler WA, Carlson TF, Jones WJ. (2004) Carbon black. *Rubber Compounding*, 239–284.

4. Donnet JB. (1994) Fifty years of research and progress on carbon black. *Carbon* **32**: 1305–1310.

5. Donnet J-B, Voet A. (1976) *Carbon Black*. Marcel Dekker Inc., New York.

6. Lin JH. (2002) Identification of the surface characteristics of carbon blacks by pyrolysis GC-MASS. *Carbon* **40**:183–187.

7. O'Reilly JM, Mosher RA. (1983) Functional groups in carbon black by FTIR spectroscopy. *Carbon* **21**: 47–51.

8. Sadezkya, Muckenhuber H, Grothe H, Niessner R, Pöschl U. (2005) Raman microspectroscopy of soot and related carbonaceous materials: spectral analysis and structural information. *Carbon* **43**: 1731–1742.

9. Takada T, Nakahara M, Kumagai H, Sanada Y. (1996) Surface modification and characterization of carbon black with oxygen plasma. *Carbon* **34**: 1087–1091.

10. Boehm HP. (2002) Surface oxides on carbon and their analysis: a critical assessment. *Carbon* **40**: 145–149.

11. Langley LA, Fairbrother DH. (2007) Effect of wet chemical treatments on the distribution of surface oxides on carbonaceous materials. *Carbon* **45**: 47–54.

12. Suarez D, Menendez JA, Fuente E, Montes-Moran MA. (1999) Contribution of pyrone-type structures of carbon basicity: An ab initio study. *Langmuir* **15**: 3897–3904.

13. Dr. Richard Hall, private communication.

14. Aldridge GR, Jaffe EE, Matrick H. (1967) Quinolonoquinolone pigments, US334102.

15. Tsuchiya K, Sato T. (2005) Aqueous pigment dispersions with vivid yellow color and good stability, and recording liquids containing them, JP200541971.

16. Shakhnovich AI. (2007) Method of preparing yellow pigments, WO2007047975.

17. Shakhnovich AI. (2006) Fluoroquinolonoquinolones and inkjet ink compositions comprising the same, WO2006102500.

18. Patton TC. (1979) Paint flow and pigment dispersion. Wiley-Interscience, New York.

19. Nagasawa H, Shimizu Y. (2006) Organic pigment fine particles and method of producing the same, WO2006132443.

20. Tabayashi I, Kazunari K, Inoue S, Doi R, Osawa N. (2000) Jet ink and process for preparing dispersion of colored fine particles for jet ink, US6074467.

21. Ma S-H, Ford C. (1999) Block copolymers of oxazolines and oxazines as pigment dispersants and their use in ink jets, EP0915138.

22. Williams DS, Carroll JB, Shakhnovich AI. (2007) Inkjet ink compositions comprising polymeric dispersants having attached chromophore groups, WO2007089859.

23. Shakhnovich AI. (2007) Inkjet inks and methods of preparing the same, US7300504.

24. Nagasawa T. (1994) Water based pigment ink, EP 688836.

25. Arai H, Kono M. (2001) Carbon black pigments for water-thinned inks, JP2001081355.

26. Tsubokawa N. (1992) Functionalization of carbon black by surface grafting of polymers. *Polym Sci* **17**: 417–470.

27. Belmont JA, Galloway CP, Amici RM. (1998) Reaction of carbon black with diazonium salts, resultant carbon black products and their uses, US5851280.

28. Donnet JB, Henrich G. (1960) Reactions radicalaires et chimie superficielle du noir de carbone. *Bull Soc Chim Fr*: 1609–1618.

29. Aboutes P. (1970) Sulfonated carbon black, US3528840.

30. Bergemann K, Fanghänel E, Knackfuss B, Lüthge T, Schukat G. (2004) Modification of carbon black properties by reaction with maleic acid derivatives. *Carbon* **42**: 2338–2340.

31. Watson JW, Kendall CE, Jervis R. (1962) Modified carbon black, GB910310.

32. Srinivas B. (2004) Surface modification of carbonaceous materials by introduction of gamma keto carboxyl containing functional groups, US6831194.

33. Allongue P, Delamar M, Desbat B, Fagebaume O, Hitmi R, Pinson J, Savéant J-M. (1997) Convalent modification of carbon surfaces by aryl radicals generated from the electrochemical reduction of diazonium salts. *J Am Chem Soc* **119**: 210–207.

34. Dyke CA, Stewart MP, Maya F, Tour, JM. (2004) Diazonium-based functionalization of carbon nanotubes: XPS and GC-MS analysis and mechanistic implications. *Synlett*: 155–160.

35. Belmont JA, Johnson JE, Adams CE. (1996) Ink jet ink formulations containing carbon black products, US5571311.

36. Gu F, Belmont JA, Palumbo PS, Corden BB, Yu Y, Halim E, Burns EG. (2007) Modified colorants and inkjet ink compositions comprising modified colorants, WO2007053564.

37. Palumbo PS, Adams CE. (2004) Polymers and other groups attached to pigments and subsequent reactions, US6723783.

38. Johnson JE, Bian N, Galloway CP. (2002) Modified pigments having improved dispersing properties, US6478863.

39. Yu Y. (2006) Pigment surface modification via nucleophilic treating agents. IS&T's NIP22: International Conference on Digital Printing Technologies, pp. 197–200.

40. Shakhnovich AI. (2007) Modified organic colorants and dispersions, and methods for their prepartion, EP1620510.

41. Miyabayashi T. (2005) Microcapsulated pigments, their manufacture, their storage-stable aqueous dispersions, and their anticlogging jet-printing inks giving images with good durability and appearance, JP2005120136.

42. Reipen T, Plueg C, Weber J. (2007) Pigment concentrates based on dike-topyrrolopyrroles, WO2007045311.

43. Bagai SK, Topham A. (1970) Phthalocyanine pigment compositions, DE2017040.

44. Baebler F. (2001) Pigment particle growth and/or crystal phase directors, US6264733.

45. Johnson JE, Belmont JA. (1998) Colored pigment compositions and aqueous compositions containing same, US 5837045.

46. Shakhnovich AI. (2006) Dispersant chemistry gives up its secrets. Reactions in diazonium treatment differ greatly from one pigment to another. *Eur Coatings J* **6**: 28–30, 32–33.

CHAPTER 7

Formulation and Properties of Waterborne Inkjet Inks

Christian Schmid
Hewlett-Packard

INTRODUCTION

Since the introduction of HP's ThinkJet printer in 1984, to the end of 2007, over 500 million inkjet printers have been sold for home and business use by numerous manufacturers. Nearly all of these printers utilize water-based inks. In this chapter, a general overview of the composition and properties of water-based inkjet inks will be given. Particular emphasis is placed on the relationships between ink properties and print performance attributes. The unmistakable theme arises, that ink properties must be finely *balanced* to meet demanding performance specifications.

A generic inkjet ink formula is shown in Table 1 below. Though the principal ingredient is water, the "other stuff" — the co-solvents, surfactants, colorants, and other additives — lend the ink many useful properties. Hundreds of distinct ink formulations have been sold by numerous manufacturers, each tuned for a slightly different set of applications. Even if such formulations were readily disclosed to the public, it would be impractical to attempt to relate each ink additive to a particular property or performance attribute, simply

Table 1. Overview of typical inkjet ink components.

Component	Purpose
Water	Primary solvent, carrier fluid
Colorants (0.5–10%)	Produce vibrant, long-lasting images
Pigments Dyes	
Co-solvents (5–50%)	Prevent nozzles from drying out
	Retain paper sheet flatness after printing
	Enhance ink film formation
Surfactants (0–2%)	Improve wetting of ink on media
	Reduce "puddling" of ink on print head
	Reduce resistor deposits
Polymeric binders (0–10%)	Improve durability of prints
	Improve gloss of prints
Other additives Biocides	Prevent growth of microorganisms
Chelating agents	React with free metals
Anti-corrosion additives	Prevent corrosion

because such relationships are complex, and may vary from ink to ink. The approach of this chapter is to first introduce the physics and chemistry behind the print process, and discuss in general the role that ink properties play. At the end, a simple formulation will exemplify some of the general formulation principles.

FIRING INK FROM THE PRINT HEAD

Creating acceptable prints requires two steps: (1) getting ink to fire reliably from a print head onto paper, and then (2) getting ink to behave in a desirable manner on paper. In the first part of this chapter, the inkjet printing process is discussed: how is ink ejection impacted by ink properties? What are typical impediments to reliable drop ejection?

Drop Ejection Cycle

The ejection process of an ink drop from a thermal inkjet (TIJ) print head begins as an electrical pulse is applied to a resistor. Within ~2–5 microseconds, the resistor surface reaches a temperature of

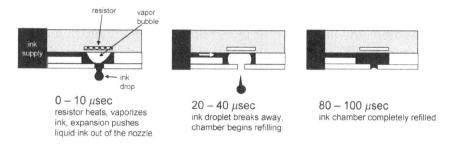

Fig. 1. Ejection cycle of a thermal inkjet (TIJ) print head.

$\sim$200–300°C, ink near the resistor surface boils, and the resulting high pressure ($\sim$100 times atmospheric) vapor bubble pushes ink through a nozzle. Ink is then drawn by capillary action from an ink reservoir to refill the chamber. See Fig. 1.

Conversely, in a piezoelectric inkjet (PIJ) head, the deflection of a membrane drives ink through each nozzle — schematics of the various configurations used in PIJ heads can be found elsewhere.[1,2] The timescale for PIJ drop ejection is similar to that in a TIJ head (Fig. 1), thus, both are capable of firing $\sim$10 000 to 30 000 drops from a nozzle each second. Typical nozzle diameters ($d = 10$–$50\,\mu$m), ink viscosities ($\eta = 1$–5 centipoise), ink surface tensions ($\sigma = 20$–$50\,$dyne/cm), and ink densities ($\rho = 0.9$–$1.1\,$g/ml) are fairly similar for the TIJ and PIJ printers for office and home use. The resulting key fluidic parameters for such print heads are summarized in Table 2.

Clearly, fluid inertia is important in drop ejection, since $Re \gg 1$; however, flow is still considered to be in the laminar regime, as the

Table 2. Typical fluidic parameters for ink drop ejection from TIJ and PIJ print heads.

Drop weight $w = 2$–$50\,$nanograms (ng)

Drop ejection velocity $v = 5$–$15\,$m/sec

Reynolds Number $\left(\dfrac{\text{fluid inertia}}{\text{viscous forces}} \right)$ $Re = \dfrac{\rho d v}{\eta} = 50$–$500$

Weber Number $\left(\dfrac{\text{fluid inertia}}{\text{surface forces}} \right)$ $We = \dfrac{\rho d v^2}{\sigma} = 20$–$300$

onset of turbulence in tube flow typically occurs for $Re \geq 2000$.[3] Furthermore, the ratio $\sqrt{We}/Re$ (also called the Ohnesorge number and represented as $\eta/\sqrt{\rho d\sigma}$) is ≤ 1 for flow in inkjet heads, thus surface forces are as important, or even more important, than viscous forces in the drop ejection process. The impact of ink viscosity and surface tension on drop weight and drop velocity has been studied both experimentally[4] and computationally.[5] Though these relationships depend in a complex way on other factors such as flow channel geometry in the print head, it is generally found that increasing ink viscosity leads to decreasing drop weight and drop velocity. In TIJ print heads, thermal energy transfer to the ink stops as soon as a vapor bubble forms, placing practical limitations on the amount of energy that can be imparted to an ink drop. Hence, TIJ heads are primarily limited to inks of low (1–5 cp) viscosities. In contrast, the mechanical displacement of the membrane in a PIJ head can be more readily extended to provide enough energy to eject inks with viscosities up to $\sim$50 cp. However, with such high viscosities, the maximum firing frequency is reduced, to accommodate the decreased rate of ink refill into the firing chamber.

Drop Break-up

Rapidly ($\sim$10–50 μsec) after ejection, an elongated column of ink breaks into distinct spherical drops, typically a large head and a few smaller satellites (Fig. 2). Satellite drops can cause problems. Because they decelerate faster under viscous air drag than the larger, main droplet,[4] satellites may not make it to the intended spot on paper in the time ($\sim$100 μsec) typically allotted. As a result, satellite drops can either end up in undesirable locations on the paper, or equally as bad, they can deposit on the internal components of the printer. Though satellite and aerosol droplet formation is complex, in general, liquid atomization studies show that smaller, finer droplets form as the Weber and Reynolds numbers increase[6] — i.e., for faster flows, and for fluids with lower ink viscosity and surface tension. Because many ink surfactants can take longer than 100 μsec to migrate to the liquid/air interface, it is possible that ink *dynamic* surface tension

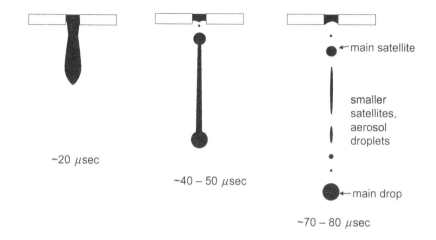

Fig. 2. Typical drop break-up dynamics from an inkjet print head.

plays a role in both drop break-up and spreading.[7] One ink formulation approach to prevent aerosol employs small amounts of high molecular weight polymers in inks, to apparently increase the ink extensional viscosity, and thus resist the formation of long liquid filaments.[8] Some PIJ heads employ subtle vibrations of the drive membrane to control the position of the meniscus at the ink/air interface, thereby helping drops break off cleanly.[9] Overall, drop break-up is a complex subject, and to fully understand it as a function of ink composition, a high speed camera will be useful.

Ink on the Orifice Plate — Puddling

Inevitably, some of the ejected ink ends up on the "orifice plate," (the outside of the print head, near the nozzles). Large ink puddles near the nozzle can rob a newly formed drop of its momentum, causing misdirection, or complete inhibition of firing. Hence, less ink on the orifice plate is desirable. While print head modifications, such as using non-wettable material for the orifice plate, or employing counterbores around nozzles,[10] can help keep the orifice plate clean, ink composition also plays a role. In particular, the use of fluorosurfactants in inks has been shown to prevent puddling.[11] The mechanism of this effect is not fully understood, but it is speculated that the fluorinated hydrophobes in such molecules prefer the ink/air interface to the ink/solid interface, thus reducing ink surface tension without promoting wetting on the orifice plate. It is also possible that a fluorosurfactant's ability to dramatically lower the ink surface tension may help suppress surface tension gradients, thus preventing Marangoni flows that would promote ink spreading out of nozzles, onto the orifice plate.

Resistor Fouling — Kogation and Decel

TIJ print heads present the complication that, over time, deposits can form on resistors, reducing the efficiency of heat transfer to the ink, and thus lowering drop weights and drop velocities over a print head's life. This phenomenon is referred to as "kogation." Though the chemistry of deposit formation is complex, one key parameter is the *solubility* of various solutes in the ink. During each firing, roughly $\sim$0.02 ng of fluid is vaporized over a resistor to generate a $\sim$20 ng ink drop. Nonvolatile ink components, in the 0.02 ng of ink, such as dyes and pigments, are left behind. If not readily redissolved or redispersed, such solutes can accumulate on resistors. The *rate* of redissolution of solutes is important — slowly redissolving components on resistors can give rise to a transient kogation-like phenomenon referred to as "decel," short for "deceleration." As the name implies, inks exhibit drop velocity decreases with firing. However, after resting for a period of seconds to minutes, the drop velocity may return to its initial value upon firing again, only to subsequently decrease again. See Fig. 3.

Furthermore, certain compounds in inks may have special affinity for resistor surfaces. For example, trace inorganic compounds are suspected to undergo electroless plating-like deposition on the tantalum oxide covering resistors, and thus the use of sequestering agents is common.[12] Ink formulation strategies to prevent resistor deposits have focused on including components in the ink that supposedly passivate the resistor surface, or slowly etch the surface.[12,13] In the latter case, a balance must be struck between stripping away unwanted deposits, and corroding resistor materials.

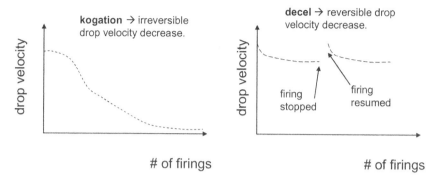

Fig. 3. The effect of resistor fouling problems (kogation and decel) on drop velocity.

Nozzle Plugging — Decap

Water evaporates rapidly from ink in inactive, uncapped nozzles. Such water loss can result in dramatic changes in the physical properties, and the phase behavior, of ink in exposed nozzles, which in turn can lead to poorly firing or completely plugged nozzles. To mitigate these effects, low-volatility "humectant" co-solvents are added to commercial inkjet inks, at levels ranging anywhere from 5 to 50% of the total ink weight. Much work goes into picking the right combination of co-solvents in inks, so that a print head can have a long "decap time" or "idle time"; that is, the head can be left inactive and uncapped for prolonged periods, but still shoot drops accurately when firing commences.

Resistance to the diffusion of water in the ink channels $\left(R_{ink} \approx \frac{L}{D_{water,ink}}\right)$ is usually ~ 100 times greater than the resistance to the transfer of water from ink to the air phase at the nozzle opening $\left(R_{air} \approx \frac{\delta_M}{D_{water,air}}\right)$, since the mass transfer boundary layer thickness between the liquid and air phase (δ_M) is usually of the same magnitude as the diffusion path length in ink channels (L), but the diffusion coefficient of water in air ($D_{water,air}$) is ~ 100 times greater than the diffusion coefficient of water in the ink ($D_{water,ink}$). As water loss from a print head is ink-diffusion limited, a substantial water concentration gradient develops in ink channels. At typical ambient relative humidities ranging from 30–80%, the equilibrium weight percent of water in exposed nozzles is estimated to range from ~5–30% respectively.[14] Furthermore, this water concentration gradient develops *rapidly* — the timescale for evaporation is $t \approx L^2/D_{water,ink}$, and for typical ink channel of length $L = 50\,\mu m$, and assuming $D_{water,ink} \approx 5 \times 10^{-6}\,cm^2/sec$, this means that the ink in the nozzle region is equilibrated with the atmosphere in as fast as 5 seconds. The situation is depicted in Fig. 4.

Certainly, dealing with water loss is essential to formulating waterborne inks for inkjet printing. It is important to understand how ink properties, such as viscosity, as well as the solubility of various components, change with water concentration. This is true for both TIJ *and* PIJ heads — because PIJ heads can be designed to eject more viscous inks, PIJ inks often employ slightly higher loadings of solutes such as pigments and polymers, and thus may exhibit a greater tendency for complications, like precipitation, as water evaporates. Furthermore, water concentration gradients in ink channels have been shown to give rise to the rapid migration of charged

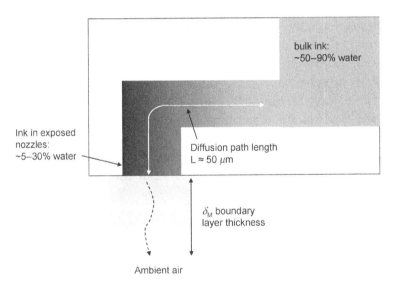

Fig. 4.   Cross-section view of the nozzle region and ink channels of an inkjet print head, showing the sharp water concentration gradient that evolves at an inactive, exposed nozzle.

pigment particles *out of* nozzles,[15,16] thus water loss can impact the composition of inks in nozzles in complicated ways. Overall, while the guiding principles above are employed to develop inks with good decap performance, ultimately some empiricism is involved.

Other Things that Block Nozzles

The narrow ($\sim$10–100 μm) ink channels in inkjet print heads can be easily obstructed. For example, air bubbles can lodge in channels, preventing ink flow. Surfactants used in the inks can stabilize bubbles, thus anti-foaming agents are sometimes added to inks to prevent excessive bubble formation/foaming. Furthermore, microorganisms can block ink channels. Thus, biocides are commonly added to inks.[12] Finally, pigmented inks add the additional complication that large ($>$500 nm) pigment particles may settle in nozzles, especially if print heads are inactive for prolonged periods. Thus, dispersions must be fine enough such that settling is not a major issue.

INK ON PAPER

Despite the numerous obstacles mentioned earlier, inkjet manufacturers are able to produce print heads that reliably put drops on paper. The next challenge to achieving good print quality is to get ink to behave properly once it is on paper. In many ways, the ink characteristics needed for effective drop ejection (low viscosity, good stability/solubility of components, poor wetting on print head exterior) are completely *opposite* those needed for good performance on paper. In this chapter, some of the basics of ink/media interactions are explored.

Drop Impact onto Paper

During the first $\sim$10 μsec of drop impact, the initially spherical drop spreads out, bulging at the edges, then recoiling. By $\sim$20–80 μsec, the drop reaches a static configuration, roughly the same diameter as the final spot size, and then begins to shrink as fluid penetrates

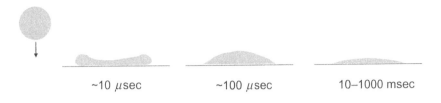

~10 μsec   ~100 μsec   10–1000 msec

Fig. 5.   Snapshots of drop impact on a solid substrate.

into paper, and evaporates.[17] Timescales for these processes depend heavily on the media type, as well as heating conditions, but for example, under ambient conditions, a ~25 ng ink drop takes ~10−50 msec to fully penetrate into porous paper.[18] In contrast, such a drop may take ~100−1000 msec to evaporate,[19] so on porous paper, most ink is absorbed before it evaporates. The ratio of final spot diameter to the original drop diameter is an important parameter, since it quantifies the covering power or "dot gain" of a drop. Experimental studies have shown that dot gain increases with increasing Reynolds number, increasing Weber number, and decreasing contact angle, with dot gains of ~1.5−3 being fairly common for inkjet drops.[17,20,21]

Optical Density

Generating vibrant images on paper is one of the primary goals of printing. A parameter used to quantify the darkness of prints is the optical density, $OD = -\log(I/I_0)$, where I_0 is the intensity of incident light on a print, and I is the intensity of light reflected from a print. Alternatively, color coordinates such as L^*, a^*, b^* are also used to quantify color prints, where L^* is the luminance (lower $L^* \to$ darker), a^* represents the red/green balance of a color, and b^* the yellow/blue balance. A related quantity, the chroma $c^* = \sqrt{a^{*2} + b^{*2}}$ represents the overall intensity of color; $c^* = 0$ indicates a neutral gray shade, high c^* indicates a brilliant color.

Both soluble dyes and dispersible pigments are employed as colorants in inkjet inks. The choice of colorant type depends somewhat on the application and media to be used. For example, on plain copier paper, light that is reflected off the surface of the ink film (without passing through the film), emanates in many directions due to the irregular paper surface. Minimizing this reflection is the key to achieving high black optical density.[22] Thus, highly-structured carbon black pigments are often used to impart a non-reflective, matte finish on plain paper. In contrast, for colors on plain paper, unwanted absorption (for example, the absorption of blue and red light by magenta colorant) is magnified with each reflection of light through the ink film.[23] Thus, dyes are sometimes preferred to provide higher chroma on plain paper, since their films typically exhibit less unwanted absorption and scattering than those of pigments.

Pores on plain paper are quite large (~1–10 microns). Thus, keeping colorants on or near the paper surface can be challenging. Various attempts at putting precipitants for colorants, such as acids, salts, and oppositely charged entities on the paper have been employed.[24] In addition, increasing ink viscosity and decreasing wetting (i.e., a higher contact angle) may help to slow ink penetration into media. Finally, simply putting more colorant in the ink may help to boost optical density. Of course, such strategies must be balanced, respectively, by the cost of media pretreatment, the deleterious effects of slow ink penetration (increased dry time, worse bleed), and the increased challenge to reliably eject inks with high solute levels.

On glossy media, such as photo paper, surface reflection can be focused away from the eye, thus glossy prints typically have higher OD and chroma than those on plain paper. Inks must be formulated to form smooth films on glossy media, and this often requires that dyes remain soluble, or pigments remain stable, in the ink as water evaporates. Gloss uniformity, across colors and ink levels, is also important on glossy papers. Achieving good gloss and gloss uniformity is perhaps more conveniently done with dyes. However pigment dispersions with controlled particle size distributions also seem to yield excellent gloss performance.

Image Permanence — Light Fastness

Over time, with exposure to light and gas phase reactants (oxygen, ozone, etc.), inkjet colorants may undergo photochemical reactions on paper. This causes original colors to fade, and original hues to shift. Pigments typically provide better image permanence than dyes, since chromophores are bundled together, providing some measure of protection against harmful reactants, and making it easier to dissipate the energy of absorbed photons. Much work has gone into making dyes that fade more slowly, for example, by incorporating transition metal complexes into chromophores.[25] The outputs of current inkjet printers have been shown to withstand substantial fading for the equivalent of one hundred years or more.[26]

Bleed

During the printing process, a region of one ink color may contact that of another. It is desirable that a coherent edge between colors is maintained, and the colors do not "bleed" together, especially if the darker ink flows into the lighter one. To combat bleeding, inks have often been formulated such that components in the black ink react with components in the color ink — for example, the black ink may contain a pigment dispersed with carboxylate groups, and the yellow ink may contain a few weight percent of succinic acid,[27] thus the black pigment loses its stabilization and rapidly flocculates in the presence of the yellow ink. In other cases, the black colorant has been designed to interact with the paper itself, thereby preventing colorant from spreading laterally when it hits the page.[28] Furthermore, surface tension differences between adjacent inks may give rise to surface tension gradient-driven flows (i.e., Marangoni flow), thus a bleed reduction strategy has been developed to use a less-wetting, high surface tension black ink with lower surface tension colors.[29]

Paper Cockle and Curl

As water penetrates into paper, it swells fibers, and also disrupts hydrogen bonds that hold paper fibers together. Over a short (~0.1–10 sec) timeframe, this can lead to paper cockle, a localized warping of the original flat sheet (see Fig. 6). The simple way to fight cockle is to put less water on the sheet. Over longer (1 minute

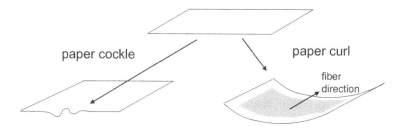

Fig. 6. Typical paper deformation induced by water-based inks.

to 1 week) time frames, the co-solvents used in inks may also evaporate. Typically, the solvent on the outsides of fibers evaporates first, allowing hydrogen bounds between fibers to reform, while the solvent *inside* fibers remains. As the intrafiber solvent slowly evaporates, fibers shrink, but cannot slide past each other, and thus sheet deformation occurs. In particular, if one side of a page is printed heavily, and the other unprinted, this deformation manifests itself as a large scale curling on the paper, as shown above. It has been found that using ink co-solvents of extremely low volatility can dramatically reduce curl over practical timescales (weeks to months).[30]

Water Fastness and Abrasion Durability

Due to the soluble nature of inkjet colorants, especially dyes, inkjet prints traditionally have been easily damaged by water. Recent advances in dyes have made them harder to redissolve,[31] and furthermore, dye "fixing" agents have recently been employed to bolster water fastness.[32] Pigments in general are more resistant to water damage, however, keeping the surface charge on pigments moderately low has been shown to lead to superior water fastness. Abrasion resistance has also traditionally been a challenge for inkjet inks, especially pigment-based inks. The inclusion of solution and dispersed polymers into inkjet inks has helped to improve rub resistance and highlighter smear.[33,34] Recently, approaches at encapsulating inkjet colorants with polymers have attracted attention.[35]

EXAMPLE — BLACK INK FOR PLAIN PAPER

We end this chapter with an example that illustrates some of the aforementioned concepts. The inks described here are completely fictitious — physical properties are estimated and not measured — however, attempts are made to include ink additives from actual patent examples.

We consider the case of a black ink for plain paper, to be applied by TIJ print head. As mentioned before, black pigments are quite effective at achieving high optical density on porous media, so we pursue a carbon black dispersion as the colorant. An investigation of the patent literature shows that a charge stabilization level of ~1.5 meq per gram pigment seems to provide a decent balance of jettability/decel/kogation and water fastness. Furthermore, choosing carboxylate groups to disperse the black pigment will enable the dispersion to crash in the presence of acidic color inks, thus providing good bleed control, and allowing under-printing with color inks to boost black optical density. Thus, we choose a suitable black dispersion, for example Cabojet 300 from the Cabot Corporation, at a level of 5% in the ink. Furthermore, we add the anionic binder SMA 2000 from Sartomer, a copolymer with 2 moles styrene per 1 mole maleic anhydride and MW = 7500, at 2% by weight in the ink to provide durability. This polymer is neutralized with sodium hydroxide, rather than the salt of volatile base (e.g., ammonium hydroxide), to allow for better pH control in exposed nozzles. Finally, the typical inkjet solvent 2-pyrrolidone (10%), and surfactant Surfynol 465 (0.1%) from Air Products, are employed. The "Revision 1" ink formulation is shown in Table 3.

In testing the Rev 1 ink, we observe excellent optical density and bleed performance, as expected. We get good water fastness, but abrasion resistance (dry rub, highlighter smear) is poor for this ink. To bolster performance, we exchange some of the styrene-acrylic binder SMA 2000 with an aqueous polyurethane dispersion, for example, Mace 85-302-1 (Mace Corporation). Furthermore, the decap

Table 3. Example of three revisions of a black ink for plain paper printing.

	Rev 1	Rev 2	Rev 3
Colorant	5% Cabojet 300	4% Cabojet 300	4% Cabojet 300
Binder	2% SMA 2000	0.7% SMA 2000	0.7% SMA 2000
		0.7% Mace 85-302-1 polyurethane	0.7% Mace 85-302-1 polyurethane
Solvents	10% 2-pyrrolidone	10% 2-pyrrolidone	10% 2-pyrrolidone
		5% tetraethylene glycol	5% tetraethylene glycol
Surfactants	0.1% Surfynol 465	0.1% Surfynol 465	0.05% Surfynol 465
			0.1% Zonyl FSA
			0.3% Phospholan 9NP
Other additives			0.05% EDTA
Viscosity	2.8	2.3	2.3
Surface tension	35	35	30
pH	8.0	8.0	8.0
Performance	• High optical density • No bleed • Decent water fastness • Poor abrasion resistance • Poor decap • Moderate kogation	• Decent optical density • No bleed • Decent water fastness • Improved abrasion resistance • Improved decap • Moderate kogation • Poor orifice plate cleanliness	• Decent optical density • No bleed • Decent water fastness • Improved abrasion resistance • Improved decap • Reduced kogation • Improved orifice plate cleanliness

performance of the Rev 1 ink is poor, so we take two approaches: reduce the overall loading of pigment dispersion and free binder, and screen for solvent combinations that lead to improved decap. The resulting formulation is seen in Table 3 as "Rev 2." Abrasion resistance and decap performance improve over Rev 1. Not surprisingly, the ink viscosity has decreased slightly — commonly lower ink viscosity tracks with better decap. To still achieve decent optical density with the decreased pigment loading in Rev 2, increased under-printing with the acidic color ink is employed.

Despite the advances made with the Rev 2 formulation, reliability aspects, such as orifice plate cleanliness, and kogation, are still poor. To address the orifice plate cleanliness issue, we add 0.1% of Zonyl FSA (an anionic fluorosurfactant from DuPont), following what has been done in the patent literature.[11] Furthermore, we add the combination of 0.3% Phospholan 9NP (a nonylphenol ethoxylate phosphate ester, from Akzo Nobel corporation), and 0.05% EDTA (ethylenediaminetetraacetic acid), to prevent resistor deposits, as discussed in the patent literature.[12] With such modifications, the Rev 3 ink gives improved orifice plate cleanliness and kogation. The surface tension is lower for this formulation. Though this can lead to bleed problems on paper, here we have specified that the color inks will be at low pH, hence will react with the black pigment, and thus combating bleed problems. Furthermore excessive tail/aerosol formation may become a problem with the decreased surface tension, as mentioned back in Section 1.2 — this performance must be checked.

In conclusion, there are many performance attributes to consider when formulating water-based inkjet inks, and multitudes of ink additives that can be employed. To achieve effective formulations, it is useful to first consider the relationship between ink physical properties and desired performance targets. Taking such an enlightened approach helps narrow the list of ink additives to be tried, expedites ink development, and ultimately increases a product's chances for commercial success.

ACKNOWLEDGMENT

The author wishes to thank Dr. Paul Bruinsma, Dr. Alexey Kabalnov, Dr. Zeying Ma, and Dr. Satya Prakash for their helpful discussions.

REFERENCES

1. Zhang J. (2005) New developments in Epson's inkjet head technology. In *Proc IS&T's 21ˢᵗ Int Conf Digital Printing*, September, 2005, Baltimore, MD, pp. 269–272.
2. Le H. (1998) Progress and trends in inkjet printing technology — part 2. *J Imag Sci Tech* **42**(1): 49–64.
3. Bird RB, Stewart WE, Lightfoot EN. (1960) *Transport Phenomena*, Chapter 6. John E. Wiley & Sons, New York.
4. Dong H, Carr W, Morris J. (2006) An experimental study of drop-on-demand drop formation. *Phys Fluids* **18**: article 072102.
5. Chen P, Chen W, Ding P, Chang S. (1998) Droplet formation of a thermal sideshooter inkjet printhead. *Int J Heat Fluid Flow* **19**: 382–390.
6. Ibrahim E, Przekwas A. (1991) Impinging jets atomization. *Phys Fluids A* **3**(12): 2981–2987.
7. Zhang X, Basaran O. (1997) Dynamic surface tension effects in impact of a drop with a solid surface. *J Colloid Interface Sci* **187**: 166–178.
8. Lee S, Webster GA, Pietrzyk JR, Barmaki, F. (2004) Hewlett-Packard, US6790268 B.
9. Epson literature: http://www.epson.com and search on 'meniscus control'.
10. Courian K, Agarwal A. (2003) Hewlett-Packard, US patent 6527370.
11. Ma Z, Anderson R. (2006) Hewlett-Packard, US patent 7129284 B2.
12. Bruinsma PJ, Sader R, Chatterjee AK, Doumaux HA, Lassar NC, Giere MD. (2003) Hewlett-Packard, US patent 6610129B1.
13. Halko D. (1991) Hewlett-Packard, US patent 5062892.
14. Internal mass transfer modeling studies at Hewlett-Packard.
15. Thakkar S, Sun J. (2003) Lexmark,US Patent 6585818 B2.
16. Kabalnov A, Wennerstrom H. (2006) Hewlett-Packard, US Patent 20060162612 A1.
17. Asai A, Shioya M, Hirasawa S, Okazaki, T. (1993) Impact of an ink drop on paper. *J Imag Sci Tech* **37**(2): 205–207.
18. Yip K, Lubinsky A, Perchak D, Ng K. (2002) Measurement and modeling of drop absorption time for various ink-receiver systems. In Proc *IS&T's NIP18: 2002 Int Conf Digital Printing Technologies*, pp. 378–382.

19. Fukai J, Ishizuka H, Sakai Y, Kaneda M, Morita M, Takahara, A. (2006) Effects of droplet size and solute concentration on drying process of polymer solution droplets deposited on homogeneous surfaces. *Int J Heat Mass Transfer* **49**: 3561–3567.

20. Ok H, Park H, Carr W, Morris J, Zhu J. (2004) Particle-laden drop impacting on solid surfaces. *J Dispers Sci Technol* **25**(4): 449–456.

21. Kannangara D, Zhang H, Shen, W. (2006) Liquid-paper interactions during liquid drop impact and recoil on paper surfaces. *Colloids Surf A Physiochem Eng Asp* **280**: 203–215.

22. Leekley RM, Tyler RF, Hultman JD. (1978) Effect of paper on color quality of prints. *Tappi J* **61**(10): 108–111.

23. Hunt RWG. (1995) *The Reproduction of Colour*, 5th ed, Ch. 13, pp. 264–279, Fountain Press, London.

24. Lee S, Byers GW, Kabalnov AS, Kowalski MH, Chatterjee AK, Prasad KA, Schut DM. (2006) Hewlett-Packard, US Patent 7066590 B2.

25. Dodge T, Uhlir Tsang LC, Lauw HP, Thornberry M. (2007) Hewlett-Packard, US Patent 7247195 B2.

26. See for example recent studies by Wilhelm Imaging Research, http://www.wilhelm-research.com/

27. Adamic RJ, Shields JP, Kowalski MJ. (1998) Hewlett-Packard, US Patent 5785743 A.

28. Iu K, Parazak DP, Guo D, Chen X, Doumaux HA. (2007) Hewlett-Packard, US Patent Application 20070076071 A1.

29. Stoffel JL, Prasad KA, Askeland RA, Shepard ME, Drogo F, Slevin L, Hickman MS, Holstun CL. (1999) Hewlett-Packard, US Patent 5880758 A.

30. Byers GW, Tran H. (2006) Hewlett-Packard, US Patent Application US20060233975 A1.

31. Deardurff L. (2007) Hewlett-Packard, US Patent 7156907 B2.

32. Tsao Y. (2003) Hewlett-Packard, US Patent US6652085 B2.

33. Sarkisian G, Chen X. (2005) Hewlett-Packard, US Patent US6908185.

34. Vincent K, Ganapathiappan S. (2008) Hewlett-Packard, US Patent 7371273 B2.

35. Vincent K, Ganapathiappan S. (2006) Hewlett-Packard, US Patent 7119133 B2.

CHAPTER 8

Solvent-Based Inkjet Inks

Josh Samuel and Paul Edwards

Jetrion Industrial Inkjet Systems, EFI

Solvent inks, interpreted literally, encompass a wide variety of inkjet inks. When speaking of solvent inks in this chapter, we will limit ourselves to inks based on organic solvents.

This definition precludes aqueous inks, but further distinctions remain. There are solvent inks where the vehicle is a fluid with very low vapor pressure at room temperature. These inks are based either on glycols, or on oils and are used on absorbent substrates such as paper. The inks most commonly referred to as solvent inks are those in which the carrier is a solvent that evaporates or is driven off subsequent to printing.

The role of the solvent in inkjet ink is a vehicle with which to deliver a functional material to the surface of a substrate. The solvent is normally driven off by either passive drying or an active drying mechanism.

A further distinction, when referring to inkjet inks, is the printing mechanism. Solvent inks are used widely in drop-on-demand piezo inkjet printing (DOD PIJ) and in continuous inkjet (CIJ). To a lesser extent there is an effort to introduce solvents into thermal inkjet drop-on-demand (DOD TIJ).

These different printing mechanisms have specific requirements, and hence lead to very different formulations.

FORMULATING SOLVENT INKS FOR CONTINUOUS INKJET PRINTERS

Introduction

For an in-depth understanding of the formulation of solvent inks for CIJ printers, we need to look at how CIJ technology works, the product applications and the history of ink formulation within the CIJ technology area. Although there are a few aqueous formulations used in CIJ printers, the vast majority of inks are indeed solvent-based, even if they are solvent/water or solvent/UV inks. Thus, solvent inks form the backbone of this industry and there are literally hundreds of unique solvent-based ink formulations being sold and used each day.

CIJ Technology and Ink Formulation

Continuous inkjet technology has been available in various forms for over 30 years. The earliest developments of the technology were targeted for use as plotting printers. There are really two distinct applications of the technology, single jet deflection and multijet binary printers, but one key aspect of the technology is common to both. The printers have a recirculating ink system and this has a significant impact on the ink formulation, with regards to the volatility of solvents which can be used in the inks. Indeed, it is possible to use solvents with very high volatilities, such as acetone and ethyl acetate in a CIJ system, without having problems related to nozzle clogging or blockages.

In both the single jet deflection and the binary systems, the system creates a pressurized and modulated ink stream, which forms a stream of droplets after exiting the print head nozzles. The difference in the two technologies being that in the single jet technology, the printing droplets are then deflected such that they land on the substrate in a pattern (raster) which creates a number or line or letter etc. With the binary array system, there are many nozzles firing streams of droplets and the droplets which form the image on the paper are not deflected, whereas the remaining droplets are deflected into a gutter and returned to the ink system. The important thing to note here is that in both cases the droplets are deflected via an electrostatic charge and that requires the ink formulations to have conductivity. The whole area of ink conductivity within a CIJ fluid is key to the performance and formulation of an ink.

Another important point of similarity between the various CIJ systems is that energy transfer within the ink is a critical feature relating to its printing performance. Consider the print head, which contains many components, but central to it is the drop generator and the nozzle system. Fluid is pumped via pressure through the drop generator and out via one or more nozzles. Upon exiting the nozzle, the pressurized stream of ink naturally begins to break up into droplets. To control this phenomenon to the level of accuracy required by the printer, modulation energy is transferred to the ink as it passes through and out of the drop generator. This modulation energy is created via a pulsing piezo crystal, which creates pressure pulses many thousands of times in a second. The combination of the strength of each pulse and the velocity of the ink stream provides for a precisely controllable break up of the ink stream into droplets of a given size. After the ink droplets are broken up, they are individually charged and deflected to place them on the substrate correctly.

Therefore, the rate at which the energy is transmitted through the fluid has a great impact on how the droplets are formed, both in terms of how soon after the ink stream leaves the nozzle and in terms of how large the droplets are when they form. Both of these aspects are critical to the functionality of the system. If the droplets form at the wrong time within the charge electrode, the charge may be incorrect, leading to an inaccurate drop trajectory. Also, if the droplet is the wrong size and therefore wrong mass, the trajectory will again be incorrect, leading to printing issues and droplets hitting the substrate in the wrong position. Therefore, the composition of the ink, as it relates to the way in which energy travels through the ink, is critical and must be considered when formulating.

Early Ink Formulation History

As with all new printing technologies, the earliest solvent ink formulations for CIJ printing systems were based upon previous ink technology and multiple new iterations of previous ink technologies, which would function correctly in the printing equipment. It can be seen that the vast majority of the original CIJ ink formulations were based upon a very simple formulation. The resin system comprising: Nitrocellulose, the solvent system comprising mostly Methyl Ethyl Ketone (MEK or butanone), the colorant comprising a black dye, with a chromium counter ion (Solvent Black 29, etc).[1]

These early formulations have definitely withstood the test of time and now, 30 years later, they are still widely used and probably still account for close to 45% of the total volumes manufactured per year.

So why have these early formulations and their components lasted so long? Well, there are both component and system answers to this question:

- *Components*

  - MEK is a very versatile solvent, it has a fast dry time, it dissolves many resins and dyes and it is very tolerant of high humidity conditions.
  - Nitrocellulose — This is a very inexpensive polymer, which adheres to many substrates and has good durability.
  - Solvent Black 29 — This is a very durable dye, with regards to environment (light-fastness, water solubility, etc) and has a strong extinction coefficient.

- *System*

  - It should be noted that the printing systems were designed around this ink type and all of its characteristics. This becomes important as you consider the relationship between formulation and CIJ technology. For instance, a measure of an ink's ability to transmit modulation energy is the velocity of sound in the ink. The velocity of sound is affected by formulation variables, such as solvent type, ink density, and ink solids content. Also, the ability for a droplet to cleanly break away from an ink stream and form correctly is related to a number of factors, which include the conformance of the polymer in the ink formulation. Thus, many of the system parameters are tuned to expect certain ink formulation characteristics. Some of these parameters are reasonably adjustable as you change ink types (such as drop velocity and charge), but others are harder to adjust, such as tuning the Piezo frequency, adjusting the parameters of the print raster or changing the acoustic function of the drop generator.

In summary, the early ink formulations worked well as they did have good performance characteristics with regards to end user properties (dry time, adhesion, humidity tolerance, etc). But a very significant factor is that they performed very robustly in the printer, due to the fact that the printer was designed around the formulation characteristics. When an ink works well in the field, customers, service people and sales people are all confident to use it.

CIJ Application Areas

CIJ printing technology found its home in the world of product identification, where the legislative requirements to have product and date coding drove a huge growth in the sales of these printing systems. CIJ printing technology is therefore most commonly used to date, adding sequential codes and well as 1D and 2D bar codes to a massive variety of products. These can range from cans, bottles, wire, cartons to printing onto pharmaceuticals and food. There have also been substantial applications in the world of commercial printing, where variable information is printed onto a variety of books, leaflets and plastic covers.

The first key thing to note about the applications is that the vast majority of applications involves printing onto stocks which have little or no absorbency and secondly, the printing process is usually very fast, both of these factors require the inks to dry quickly. Hence for the ink formulations, we need to take these into account and in addition, the drying speed drives the use of very volatile solvents as well.

The substantial number of different substrates and related applications is also a key factor in the ink formulation, and is a significant reason why there are so many different solvent-based CIJ ink formulations.

Other application related factors, which have an impact on the ink formulation include:

- *Solvent Smell Tolerance* — The most commonly used solvent, MEK, has a strong odor when significant printing is carried out. Solvent is not only emitted from the printed ink, but also from the jet stream and from the printer itself. This can be a problem in various workplaces or in certain environments. In order to reduce this odor and still maintain reasonable dry times, solvents such as Alcohols or Acetone or mixtures thereof can be used.
- *VOC Requirements* — The minimization of VOC's (Volatile Organic Compounds) in the workplace is another trend which has emerged in recent years. Choice of solvent mixture can again benefit this requirement, with additions of solvents such as Acetone and higher boiling point Lactate Esters.
- *Perceived or Real Health and Safety Issues* — There are a number of issues or perceived issues which customers encounter with the original MEK-based inks. The use of MEK has its share of concerns, based upon the smell and perception that it is a harmful material. Indeed, for this reason inks have been formulated with alternative solvents such as ethanol and methanol, as well as previously discussed acetone, and a variety of acetates, such as ethyl acetate. Some of these solvents have unpleasant odour, some more pleasant smell; some have a lower toxicity level. That said, methanol is commonly used as a low odor replacement, but could probably be considered a more toxic ingredient! Another component which is often replaced is the Solvent Black 29 dye, this dye does contain chromium as the counter ion and the toxicity concerns of having free Chromium 6 ions in an ink can be an issue for some applications or customers. Dyes with alternative counter ions or pigments have been chosen to replace this material. These requirements are often found in the food processing or pharmaceutical application areas, although they can occur with any customer.

- *Printing onto Colored Substrates* — Many substrates onto which coding is required are not that suitable for printing with the standard black ink or indeed other dye-based inks, as the colors are too dark for the inks to show up. In this case, the inks need to be formulated with opaque[2] colorants and these will usually be either pigments white, yellow, or blue. Sometimes the formulation will contain a mixture of white pigment for opacity and a dye to color the white.

- *Printing onto Wet Bottles* — A very large application segment requires the printing of codes onto bottles which have residual water or condensation on them from the bottling process.[3] This is a tough ink formulation application and requires some special chemistry to both cut through the water layer and adhere to the glass where water is present as a thick film. The usual solution to sticking to the wet glass surface is the application of reactive Silane chemistry, which is very effective, although the use of reactive components requires a lot of skilled formulation to ensure that the inks only react where they are intended to and not upon storage or in the printer.

- *Direct/Indirect Food Contact or Food Grade Application* — The use of solvent CIJ in food related applications is quite substantial. From printing onto the outside of packaging, to printing on areas of packaging where food will come into contact with the code, to actually printing onto the food itself, spans a very large gamut of applications. FDA and other regulatory bodies have a lot to say on which ink raw materials can be used in these applications and they are often complex and vary from country to country. The most challenging application in this area comes from printing directly onto food. One very large application area, especially in Europe, is coding onto hens eggs. Each egg has to have at least one code/date stamp to identify it. As you can imagine, this application is very restrictive to a CIJ solvent ink formulation.[4] The solvent system needs to be a mixture of pure ethanol with small amounts of water to be a suitable carrier material. The mixture not only has to dissolve the other components, but it has to act as an antimicrobial medium to stop microorganisms from growing in the ink upon storage and within the printer. The colorant has to be food approved, and in this application withstand being washed out by moisture. The usual colorant used for this application is erythrosine (a red dye) and you need a polymer which is both food approved and resistant to moisture; a small number of these polymers exist, including various cellulose-based materials. The dye itself comes in with a salt, which provides the conductivity required to achieve charge and hence deflection in the print head.

- *Bar Code/OCR Readability* — Many applications now require not only an alphanumeric code, but also a 1- or 2D or even an OCR readable code. In most cases the use of the standard black solvent ink will not suffice; therefore a number of black inks, using carbon black pigment as the colorant have been developed to achieve acceptable read rates. The carbon black has a wide range of absorbance and is very effective for use with red or infrared bar code readers. The formulation of the CIJ ink with dispersed pigments in itself often is a challenge, as CIJ inks are relatively low in viscosity, and it can be hard to stabilize the pigment in suspension. This is obviously an issue for long term storage, but can also be an issue in the printer, where there are significant sheer forces at work, especially at the nozzle. These sheer forces can be very challenging to deal with, in terms of creating a stable dispersion.
- *Fluorescence or Phosphorescence Requirements* — There are another selection of applications which require the use of fluorescent or phosphorescent colorants within them. One area where the use of these materials is quite common is in the postal industry. Many letters today have a number of bar codes, used to track, guide and trace the mail. Very often red fluorescent or phosphorescent dyes are used in the solvent ink. These dyes take in light of a given, lower wavelength and provide an emission shifted to a higher wavelength. Often these inks appear to be either pink or orange to the eye in normal light. However, in certain postal applications, such as Japan Post, there is a substantial need for a clear ink which can provide a red fluorescence. This is required so that the ink does

not negatively impact the visual quality of the cards onto which it is printed. This is actually a tough formulation challenge, as there are few materials which can provide this color shift.

These are just a few highlights of the many application areas for CIJ inks and how many different and varied formulation challenges there have been over the years. One ink may hit a great number of application areas, but many ink formulations are required to hit them all!

CIJ Ink Formulations

As we have seen, there are many technology demands and application demands placed upon the solvent CIJ ink formulator. Let us now look a little more deeply into the components of the inkjet ink formulations and how they are chosen. But first we should consider the general specifications for a CIJ ink.

There are many CIJ printers on the market today and each has its own set of physical requirements from an ink. The following is a basic list of properties into which the solvent ink formulations usually fall.

- *Viscosity* — 2 cp to 10 cp at nozzle temperature
- *Surface Tension* — 20 to 35 dynes/cm
- *Conductivity* — 700–2000 mS

As can be seen from these parameters, the inks are at low viscosity, they have a good range of possible surface tensions and the conductivity, in a solvent, is required to be quite high.

The following is an example formulation, using generic components to show typical composition and quantities:

- *Colorant* — 5%
- *Polymer* — 8%
- *Conductivity Salt (If colorant is non-conductive)* — 1%
- *Surfactants* — 0.5%
- *Carrier Solvent* — 85.5%

Let us now look more closely at each of the above categories and discuss their functions using some typical materials as examples:

Colorants

As noted in previous sections, the colorant used in the CIJ solvent ink is a key component, which can often vary from application to application. Dyes tend to be used in these formulations for a number of reasons. In some instances the dye will give a color or property which is not possible when using pigments. Dyes, especially when they provide their own conductivity via a counter ion or residual salts, are usually easier to formulate with and provide ink stability, both upon storage and upon long term use in the printer.

Care has to be taken with the purity of the dyes used. Although having associated counter ions and some free ions can be helpful in providing the conductivity you need in an ink, too much free salt or the wrong type of salt can be a problem in a number of areas. For instance, some salts are highly corrosive to some metals, and can cause problems with certain ink system wetted components, such as solenoids, which open and close valves. Another problem relates to the interaction of salts with other components in the formulation, such as the polymer or other low level raw material contaminants. There have been issues with gels forming within polymers, which can lead to serious jet movement issues. Note that jet movement to any significant level is a very serious problem with CIJ printing systems and often leads to poor printer performance and or poor print quality.

Another issue to watch out for is the dye or free salt having solubility problems in solution; at normal temperatures the materials can be very soluble, but sometimes either low or even high temperatures can lead to materials coming out of solution and clogging up the printer.

Polymers

The choice of polymer within the system is often driven by the end user requirements for the ink. For instance, if the customer wishes the ink to adhere well to glass, but also wishes the ink not to have a smelly solvent such as MEK, then the available choice of polymers which are both soluble in available solvents and also adhere to the substrate can be quite small. On top of this we have to consider the properties of the polymer in relation to the printer functionality. For instance, if the polymer has too low a molecular weight, then you may have to put vast amounts of polymer into an ink to achieve a usable viscosity. Having too much polymer has the potential drawbacks of increasing the dry time of the ink and changing the velocity of sound within the ink. At the other end of the spectrum, using a polymer with high molecular weight or viscosity, can have a negative impact on drop break off characteristics, often forming unwanted satellites which accompany the drops and affect both print quality and printer functionality.

Polymer conformation within the solvent system in question also has an impact on both the ink viscosity and the cleanliness of drop break up. Some polymer chains have a tendency to fold in upon themselves, thus not entangling as much with other polymer chains in solution. These tend to break off well and not form satellites, whereas other polymers can form very twisted and entangled networks in solution. These very entangled polymer chains more often lead to what is termed "stringy break-up" and usually form the unwanted satellites.

Polymers more commonly used in CIJ Solvent Inkjet formulations include, nitrocellulose, acrylics, cellulose acetate butyrate, vinyl chloride co-polymers, polyvinyl butyrate, etc.[5]

Conductivity Salts

When there is no conductivity imparted by the colorant, be it a dye or a pigment, conductivity salts need to be added to achieve the required conductivity for drop charging and deflection. The choice of, and effectiveness of the conductivity salt very much depends upon the solvent choice being used, but it is also important to consider the compatibility with the colorant and the polymer.

As mentioned previously, it is also important that the corrosiveness of the conductivity salt in the solvent system is also assessed as corrosion in organic solvents is hard to predict theoretically. Typical conductivity salts used as additives within CIJ ink formulations include lithium nitrate, potassium thiocyanate, lithium triflate etc.

Surfactants

CIJ inks need to work on numerous substrates and there is a relationship between drop spread on a substrate and print quality. The amount of drop spread depends upon a number of issues in a CIJ ink, including surface tension, viscosity, solvent evaporation rate, interaction with substrate, amount and type of polymer in the ink etc. You may wish to improve the print quality of a certain ink-substrate combination by optimally adjusting the drop spread. A common method to do this is by adding a surfactant to the ink formulation. Depending upon the chemistry of the surfactant, it will either increase or decrease drop spread, and hence is a good mechanism for tuning print quality. There are many hundreds of surfactants available, and you can use all chemical types, including anionic, cationic, non-anionic forms. Specific examples include polyoxyethylene fatty ethers and diethylhexyl sodium sulphosuccinate.

Carrier Solvent

The carrier solvent makes up the bulk of all solvent CIJ ink formulations and has a massive impact on the overall properties of that ink and which ingredients can be added to it to make up the final formulation. Solvents can be single, but are also very often mixtures of two or more. The choice of solvent/polymer mixture is very important for a number of reasons, but drying time and the ability for a polymer to let go of the solvent mixture is very important. Solvent mixtures are often used to ensure solubilization of certain key components, into a formulation with a good balance of properties. The bulk solvent has a significant impact on the velocity of sound of the ink and mixtures can be used to obtain a more optimum velocity of sound profile. Typical carrier solvents include: MEK, acetone, ethyl acetate, methanol, ethanol, methyl acetate etc.

FORMULATING SOLVENT INKS FOR PIEZO DROP-ON-DEMAND PRINT HEADS

Piezo drop-on-demand print head's principle of operation is relatively simple in comparison to CIJ. The jetted fluid is held in balance between a slight negative pressure in the ink system, and capillary tension in small nozzles. Each nozzle is attached to a small chamber from which ink is ejected on demand when a piezo crystal coupled to that chamber flexes on application of voltage. While the principle of operation is simple, producing a viable print head is not.[6] The print heads themselves are complex constructions and have specific requirements on the fluid to enable reliable operation.

Formulation Requirements

a. The ink must be compatible with the printer, i.e., the solvent should not damage the print head or the ink delivery system.
b. The ink should have physical characteristics that are required for the print head. Both viscosity and surface tension influence drop ejection and these requirements are head specific. Many of the Piezo print heads used today (i.e., offerings from Xaar, Fuji-Dimatix) require viscosities in the 10–14 cP range at jetting temperatures. Jetting temperature is often elevated above room temperature. This has the advantage that the jetting conditions are controlled by the printer and not by ambient conditions. Static surface tension is required in the range of 22–36 dynes/cm.
c. The ink should be stable in the print head. This requires that pigmented inks do not aggregate and block passages and nozzles, and that the binder does not dry and irreversibly clog the print head.
d. The ink should not exhibit print latency, i.e., the ink must be able to jet normally after a quiescent period. Faster drying solvents cause an increase in viscosity at the nozzle when the nozzle has

not been activated for some period of time. The high viscosity causes the first few drops to be slow. The slow jets show up as misdirected drops giving ragged edges on the print. At one point jets will be lost and a maintenance cycle, i.e., purge and wipe is required. There are various procedures that reduce this problem such as subjetting, activating the piezo crystals below a level which causes drop ejection, as this mixes the ink near the nozzle without drop ejection. An additional procedure is spitting or exercising the jets at the end of a print cycle in a scanning print process. Even with these mechanisms, solvent choice is critical in avoiding latency.

e. The ink should have an acceptable shelf life. Normally this is a year under storage conditions at room temperature. This is often tested by holding the ink at an elevated temperature for shorter time periods. Common measures are 4 weeks at 60°C and 12 weeks at 45°C. This also includes stability to transportation environment, particularly freeze-thaw cycles.

f. The ink should have good adhesion to target substrates.

g. The ink should dry quickly without having to invest large amounts of energy. This requirement is both from the perspective of invested energy but also the footprint of the printer and tradeoff with speed that would enable drying.

h. The ink should resist blocking, i.e., prints should not stick to each other when rolled or stacked.

i. For graphic arts applications the inks should have a large color gamut and, particularly for outdoor applications, should resist fading.

j. There are other specific requirements, depending on the application, such as flexibility and chemical resistance.

Formulation

Solvents

The solvent makes up the bulk of the ink. Many of the requirements listed above make conflicting demands. Choosing the correct mix of solvents is a balancing act requiring compromise to manage these conflicting requirements.

Solvents are chosen from groups commonly used in the coatings industry: ethylene glycol ethers (Cellosolve), propylene glycol ethers (Dowanol) and esters of these ethers. Ketones are common, particularly higher boiling ketones such as cyclohexanone and isophorone. Alkyl lactates are becoming increasingly prevalent because of environmental and health concerns that apply to conventional solvents.

One of the most important determining factors in the choice of solvents is the balance between the need to dry on the substrate but not suffer from latency and blocked nozzles. Using very slow drying solvents would lead to more reliable jetting, because it would minimize drying at the nozzle and hence nozzle blockage. On the other hand, slow solvents are just that, they require larger amounts of energy and longer time to be removed.

A mixture of solvents is chosen that are good solvents for the resin, with attention to the evaporation rate of the solvent. A common measure for evaporation rate is a dimensionless ratio versus the evaporation rate of n-butyl acetate (ER). Most solvents used will have an ER in the range of medium (0.8–3) or slow (< 0.3 on a scale where n-butyl acetate =1).

For example, an ink may contain a mixture of cyclohexanone with an evaporation rate of 0.2 and butoxyethanol (Butyl Cellosolve) with an evaporation rate of 0.07. Additional factors to take into account when choosing the solvents are the interaction of the solvent with the substrate, e.g., swelling of polymeric substrates enhances adhesion. Surface tension will influence print quality and dot gain with high surface tension solvents leading to smaller drops.

Colorants

While dyes have been used in the past, the colorants most commonly used in DOD inkjet are highly dispersed pigments. The pigments have the advantage of light fastness compared to dyes. Pigments are media milled to a particle size below 1 micron. Convenient sources of pigments are predispersed pigments in a chip form, for example as a 50% dispersion in solid in a vinyl chloride/vinyl acetate co-polymers. These chips are available from companies such as Clairient and Ciba. The use of chips avoids the need to mill a pigment base and makes production with simple milling equipment possible. The overall pigment content in the ink depends on pigment color and type and is in the range of 2–10%.

Binder

A resin or binder is required to impart adhesion, and physical and chemical resistance to the image so that it will not scratch or wipe off. The resin may also be used to build up viscosity for those print heads that require a higher jetting viscosity. In order for the print head to function reliably, the resin should be able to redissolve in the ink. If the resin is no longer soluble in the ink when dried, there is a large risk of plugged nozzles. This requirement must be balanced with the requirement to make the printed surface scratch and chemical resistant.

For this reason formulators usually use polymer systems that cure by evaporation and not by reactive cross linking. Cross linked systems have been used[7] but care must be taken to have a mechanism where the curing is inhibited in the print head and takes place only after jetting.

Jetting reliability also limits the molecular weight of the binder used. In the vicinity of the nozzle, the more volatile components of ink evaporate. When the resin drops out of solution this will cause nozzle blockage. In addition, high molecular weight polymers may lead to viscosities outside the print head range. Polymers used normally have a molecular weight below 100 000 and often below 50 000. Common polymers used are vinyl chloride/vinyl acetate co-polymers, acrylic resins and polyketone resins.

Additives

Additives are components added in small amounts for a variety of purposes. Surfactants such as siloxanes and flouro surfactants reduce surface tension and control dot gain. Acrylic leveling agents are used to attain uniformity across the print without greatly reducing surface tension.

An additional class of additives is for protection of the print when exposed to light or heat. Ultraviolet light absorbing materials such as substituted benzophenones help to protect both the binder and some types of pigment by absorbing higher energy light that causes degradation. Radical scavenging additives, commonly hindered amine light stabilizers (HALS) are added in small amounts both to protect the pigments and binders from weathering. Additional additives may be used depending on specific requirements. For example, plasticizers may be added to reduce cracking of films when stretching or bending of the film is required such as in vehicle wraps.

Challenges and Trends in Solvent Inkjet Inks

Radiation curable inks have been making inroads into markets where solvent inks reigned in the past, particularly wide format printing. UV and EB curable inks have the advantage of low emissions. In addition, these inks have an advantage over solvent inks in that they can be cross linked on the substrate, leading to durable prints, while the inks in the printer can be kept stable.

Solvents have the advantage that they serve as a relatively inexpensive means with which to deliver a functional material to a substrate. One of the major drawbacks in the use of solvent inks are environmental issues. The environmental issues are both workplace related: lists of volatile organic compounds that have restricted concentrations in the workplace, and broader concerns as expressed in regulations such as the Hazardous Air Pollutants (HAPS) list that limits the amounts of specific compounds that can be emitted by an operation. In response to these concerns, many inks are now marketed with an environmental twist.

Inkjet is a commercially driven endeavor, and as such it is often difficult to disentangle the marketing hype from the substance. There are numerous inkjet inks of types described by names meant to indicate that they are healthier to use or have a more benign impact on the environment. Ecosolvents, soft solvents, mild solvents and biosolvents are a partial list of phrases used to describe solvent-based inks that are ostensibly more environmental friendly than conventional solvent inks. What most of these inks have in common is that they use solvents that are less aggressive than what is commonly termed full solvent inks. Particularly they do not use ketones such as cyclohexanone and aggressive solvents such as N-methyl pyrolidone.[8] In place of these solvents, these inks commonly use alcohols, glycol ethers and lactates. The more aggressive solvents mentioned above are very effective in softening common substrates such a vinyl and enhancing adhesion. In many cases the performance of inks using the less aggressive solvents were found to have inferior adhesion and require more aggressive drying.

An additional approach to environmentally responsible inks is to source the raw materials from renewable resources. In this approach solvents are based mainly on alkyl lactates and binders such as cellulose or nitrocellulose.[9]

In the final analysis, the separation between the different types of inkjet: solvent, UV, is somewhat arbitrary and indeed there are advantages to mixing solvent into radiation curable inks.[10]

REFERENCES

1. Lechaheb *et al.* (2000) Ink for continuous inkjet printing, US patent 6106600.
2. Zhu *et al.* (2007) Opaque ink jet composition, US patent 7279511.
3. Deng *et al.* (1997) Wet surface marking inkjet, US patemt 5652286.
4. Crocker *et al.* (1994) Inks GB 2277094.
5. Burr *et al.* (1999) Ink jet inks, US patent 5998502.
6. Pond, Stephem F. (2000) *Piezo DOD Inkjet in Inkjet Technology and Product Development Stratagies*, pp 104–114. Torrey Pines Research, Carlsbad California.
7. Zou *et al.* (2000) Reactive jet ink compositions, US patent 6140391.
8. King Ke Yung *et al.* (2008) Environmental-friendly and solvent-based inkjet ink composition, US Patent Application 20080006175, January 2008.

9. Beck Charles *et al.* (2007) Inkjet inks, methods for applying inkjet ink and articles printed with inkjet inks, US Patent Application 20070043145.

10. Ylitalo *et al.* (2004) Inks and other compositions incorporating limited quantities of solvent advantageously used in inkjet, US Patent 6,730714.

CHAPTER 9

Formulating UV Curable Inkjet Inks

Sara E. Edison

Hexion Specialty Chemicals

INTRODUCTION

Inkjet printing is enjoying a huge surge of interest, due in part to advancements in hardware technology that have improved aspects such as print speed, reliability, and resolution. Photographic quality images are possible at fairly rapid print speeds and the amount of maintenance required for the printers is relatively minimal. These improvements, coupled with the inherent advantages of digital inkjet printing (such as customization of each print, cost savings for short run prints, and process savings due to the elimination of clean up and set up steps) have resulted in an expansion of inkjet printing into many markets, from graphic arts to industrial arenas.

Several types of fluids may be jetted via inkjet heads, which can be of the continuous inkjet variety (CIJ, primarily aqueous or solvent-based), thermal inkjet (TIJ, primarily aqueous, but some solvent and UV/aqueous hybrids exist), or piezoelectric drop-on-demand (DOD, aqueous, solvent, oil, UV curable). While there are advantages and disadvantages for each type of hardware and fluid, a clear leader in growth has emerged from the list. UV curable inkjet inks, applied via piezoelectric DOD print heads, are experiencing double digit growth. This is due to the many advantages that UV curable inkjet inks bring to the process as well as to the end product, such as:[1]

- Environmentally friendly — little to no volatile organic chemicals (VOCs) or hazardous air pollutants (HAPs).
- Cures instantly upon exposure to UV radiation.
- Curing lamps have a smaller equipment footprint than conventional dryers.
- Curing lamps consume less energy than conventional dryers.
- Longer open air time (time between printing that does not require an ink purge or prime to restart) resulting in less ink waste.
- More consistent ink quality as no solvent evaporation occurs (also less ink waste).
- Finished prints are durable and abrasion resistant due to the cross linked nature of the film.

With all of these benefits, it is no surprise that many print shops are making the conversion to UV curable inks for their digital processes.

There are several limitations on the physical properties of UV curable inkjet inks for piezoelectric DOD print heads. The viscosity needs to be fairly low — 8–12 cps at the jetting temperature (most print heads have on-board heaters, capable of reaching up to 70°C in many instances). The surface tension of the inks is also important, and depends on the print head technology being utilized. Some print heads are equipped with a non-wetting faceplate, while others work best with fairly high surface tension fluids (mid 20s to upper 30s, dynes/cm).[2]

These requirements make formulating UV curable inkjet inks more challenging than for applications where low viscosity is not mandatory. This chapter hopes to provide an introductory look at formulating UV curable inkjet inks for printing via DOD print heads.

FORMULATING UV CURABLE INKJET INKS

Most UV curable inkjet formulas are comprised of the following materials:

- Monomers/oligomers
- Colorants (pigments or dyes dispersed or dissolved in a reactive carrier)
- Photoinitiators
- Additives.

Reactive monomers and oligomers make up the base of an inkjet formula, giving the fluid most of its properties. These materials must possess functionality to allow free radical additions to occur.

Free Radical Polymerization Mechanisms

There are essentially two different mechanisms that UV curing may occur by — free radical or cationic. Free radical polymerization is the most predominantly used route and will be discussed first. The chain reaction that occurs consists of at least four steps:[3]

1. Initiator radical formation
2. Initiation
3. Propagation
4. Termination.

In step one, an initiator forms a free radical when energy, in this case UV radiation, is applied to the formulation (this will be described further in the next section). Next, this free radical reacts with an available monomer (or oligomer) to create a propagation site and initiate the polymerization chain reaction. The radical containing monomer (or oligomer) will continue to propagate the reaction until the monomer supply is depleted or until a termination step occurs. Termination may take place one of several ways. Two chains may combine at their propagation sites and end the reaction. Alternatively, disproportionation may occur via hydrogen atom abstraction. For example, one chain may be oxidized to an alkene and one chain

Initiator Radical Formation

Ro-OR $\longrightarrow$ 2RO•

Chain Initiation

RO•+CH$_2$=CH$_2$ $\longrightarrow$ ROCH$_2$C•H$_2$

Chain Propagating Steps

ROCH$_2$C•H$_2$+CH$_2$=CH$_2$ $\longrightarrow$ ROCH$_2$CH$_2$C•H$_2$

Termination (Two Propagating Sites Combine)

ROCH$_2$CH$_2$C•H$_2$+$_2$H•CH$_2$CH$_2$COR $\longrightarrow$ ROCH$_2$CH$_2$CH$_2$CH$_2$CH$_2$CH$_2$OR

Termination (Disproportionation, H-atom Transfer)

ROCH$_2$CH$_2$C•H$_2$+$_2$H•CH$_2$CH$_2$COR $\longrightarrow$ ROCH$_2$CH$_2$CH$_3$+$_2$HC=CHCH$_2$OR

Chain Transfer

ROCH$_2$CH$_2$C•H$_2$+XY $\longrightarrow$ ROCH$_2$CH$_2$CH$_2$X+Y•

Fig. 1.   Free radical UV curing mechanism.

may be reduced to an alkane. Finally, it is possible for a chain transfer reaction to occur, whereby the growing chain reacts with a molecule XY in such a fashion that X terminates the chain and Y* proceeds to initiate a new chain. XY can be a solvent, a radical initiator, or any molecule with a bond that is susceptible to homolytic cleavage. Figure 1 shows simplistic examples of the steps involved in the free radical curing mechanism. Step one does not have to contain a peroxide unit.

Monomer Selections

The majority of the monomers and oligomers used in UV curable inkjet ink formulations are acrylates of varying functionalities, although occasionally materials such as unsaturated polyester resins are used as well. Figure 2 shows examples of the acrylate moiety. Acrylates are skin sensitizers and should be handled with caution.

Acrylic Acid   **Acrylate Esters**

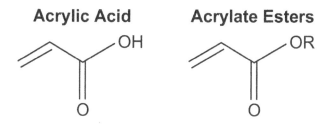

Fig. 2. Examples of acrylate functionalities.

Increasing the functionality of the acrylate will increase the cross link density of the polymer film, resulting in a harder, less flexible film that has a higher resistance to solvents, abrasion, and scratches. Also, a higher cross link density can cause the film to shrink upon curing, lowering the adhesion. Alternatively, monomers that are mono- or di-functional are more flexible but less durable, and experience less shrinkage. Ultimately, the end use of the print must be considered when choosing the most suitable monomer/oligomer blend.

One property that almost every ink must possess is adhesion to whatever media it is being printed upon. In both the graphic arts and industrial printing markets, adhesion to plastics ranks high on the list of requirements.

Polar plastics such as polymethyl methacrylate (PMMA), polyvinyl chloride (PVC), nylon, and polycarbonate are generally fairly easy to adhere to. Monomers such as 2-phenoxyethylacrylate, alkoxylated phenol acrylates, and ethoxylated tetrahydrofurfuryl acrylate are all excellent choices to increase adhesion to polar plastics and are commonly found in many ink formulas. It is crucial to know what components will be present in the print head and the ink delivery system in order to test for material compatibility. Many O-rings, tubes, ink reservoirs, and filters are composed of plastics that may be degraded by the aggressive monomers used to obtain adhesion to the media. Soaking coupons of the pieces of the printer in ink at elevated temperatures and measuring the weight change and visual appearance of the coupons can identify problem formulations.

Non-polar plastics such as polytetrafluoroethylene (PTFE), polyethylene (PE), polypropylene (PP), and polystyrene (PS) are more difficult to adhere to. 1,3-butylene glycol diacrylate has shown some efficiency in promoting adhesion to these types of substrates.

Other properties that are heavily influenced by the choice of monomer include cure speed (in general higher functional monomers cure more rapidly), viscosity, and durability of the film. Table 1 lists some monomers, their viscosities, and the properties that they enhance (reprinted with permission from Sartomer).[4] It is important to note several trends on the chart. Cure speed increases with an increase in functionality (all of the recommended monomers in that column are at least trifunctional and several are tetra- or penta-functional). Viscosity also increases as the functionality of the monomer is raised (all of the low viscosity diluents are diacrylates). The adhesion promoting monomers are all di- or mono-functional. Most formulas contain several different monomers and sometimes also oligomers as there is often a balancing act that must be performed when selecting materials that will provide the required performance properties while still maintaining the correct viscosity and surface tension.

It is also worth noting that the field of UV curable inkjet inks has recently become fairly well protected in terms of intellectual property, particularly in terms of composition. Therefore, it is advisable to do a thorough patent search before devoting excessive resources to a promising formulation.

Colorants

Another important component present in inkjet inks is the colorant, typically a pigment or dye (organic pigments are used most often due to their superior lightfastness). These are often a challenge to incorporate into the system for several reasons. The pigments must be ground to very small particle sizes (<1 micron absolute) to flow through the print head nozzles, which are typically 30–50 microns wide. These nozzles can tolerate only submicron-sized particles in order to maintain optimum jetting performance and prolong the

Table 1. List of typical monomers, properties they enhance, and viscosity.[4]

Adhesion to Plastic	Cure Speed	Low Viscosity Diluent	Film Durability
1,3-butylene glycol diacrylate, 9 cps	Di-trimethylolpropane tetraacrylate, 600 cps	Hexanediol diacrylate, 9 cps	Ethoxy (3) cyclohexanol dimethanol diacrylate, 51 cps
2-(2-ethoxyethoxy) ethyl acrylate, 6 cps	Dipentaerythritol pentaacrylate, 12 000 cps	Tripropylene glycol diacrylate, 15 cps	Ethoxy (3) cyclohexanol dimethanol diacrylate, 70 cps
2-phenoxy ethyl acrylate, 12 cps	Propoxylated (3) trimethylolpropane triacrylate, 90 cps	Dipropylene glycol diacrylate, 10 cps	Propoxylated (3) cyclohexanol dimethanol diacrylate, 80 cps
Cyclic trimethylolpropane formal acrylate, 15 cps	Ethoxylated (5) pentaerythritol tetraacrylate, 150 cps	Ethoxy (3) hexanediol diacrylate, 24 cps	
Ethoxy (3) phenoxy ethyl acrylate, 24 cps	Ethoxy (6) trimethylolpropane triacrylate, 95 cps	Ethoxy (5) hexanediol diacrylate, 46 cps	
	Propoxylated (6) trimethylolpropane triacrylate, 125 cps	Propoxylated (3) hexanediol diacrylate, 23 cps	
	Propoxylated (3) glyceryl triacrylate, 95 cps	Propoxylated (2) neopentyl glycol diacrylate, 15 cps	

life of the print head, since larger particles may ricochet around the nozzle and abrade the orifice. However, agglomeration and flocculation can occur more readily at these sizes, which can lead to sedimentation and instability. Also, the smaller particle sizes can have a detrimental effect on the weathering ability of the pigment as well as the color gamut, gloss, and opacity, which are greatly influenced by particle size. It is not only important to have a small overall particle size, it is also necessary to have a narrow particle size distribution to ensure a more homogeneous fluid for consistent jetting.

Proper dispersing techniques and additives can help to alleviate this to some degree. It is recommended to use predominantly low viscosity reactive diluents along with a dispersant specifically prepared to stabilize and completely wet the organic pigment in question.

Photoinitiators

UV curable formulations also must contain photoinitiators to promote the polymerization reactions. These are either of the Type I or Type II variety. Type I photoinitiators undergo a unimolecular bond cleavage upon irradiation to yield free radicals. Type I initiators are more versatile and consequently the most widely used. The initiators have a generally higher efficiency due to their generation of free radicals via a unimolecular process as they only need to absorb light in order to generate radicals. Figure 3 shows two examples of Type I photoinitiators, α-hydroxyl ketone on the left and 2-hydroxy-2-methylpropiophenone on the right.

Type II photoinitiators undergo a bimolecular reaction where the excited state of the photoinitiator (acting as a photosensitizer)

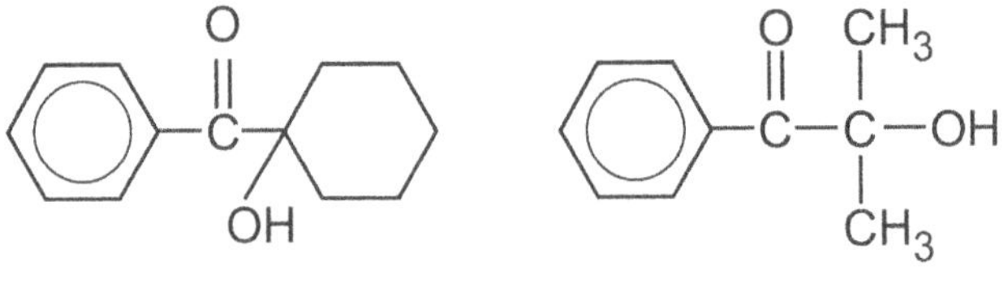

Fig. 3.   Two examples of Type I photoinitiators.

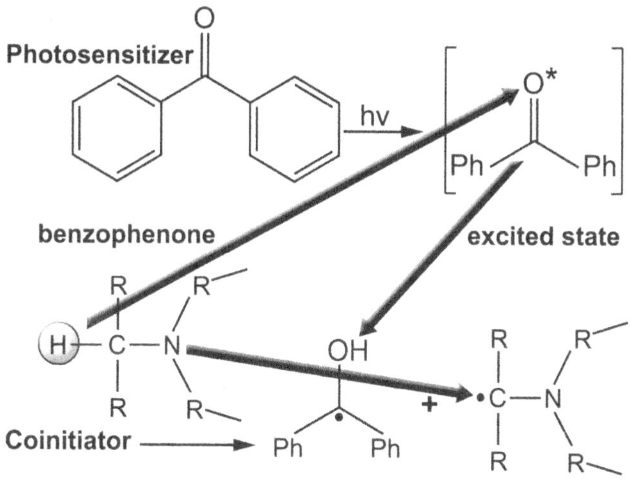

Fig. 4. Reaction of a Type II photoinitiator with a coinitiator.

interacts with a second molecule (a coinitiator) to generate free radicals. Figure 4 demonstrates the process in which a Type II photoinitiator operates.

It is important to be aware of potential acid/base interactions that can occur due to the presence of an amine synergist coinitiator. Often, pigments are surface treated or dispersed with acidic or basic materials to help prevent flocculation, and this could be detrimental to the shelf life of the ink if there is a reaction with the coinitiator. When choosing a photoinitiator is it very important to make sure that the maximum absorbance matches the wavelength that the UV lamp emits and that it is not overlapped by other components in the formula, particularly the pigment. Generally, most inks will contain several different photoinitiators to ensure optimum cure. The most commonly used lamps are standard mercury vapor bulbs (sometimes referred to as "H" bulbs), which have a broadband output but have more density in the lower wavelength, higher intensity regions of the spectrum. This allows them to promote superior surface cure. Another commonly used lamp is the iron additive mercury vapor lamps (sometimes referred to as "D" bulbs), which are rich in the UVA region and provide excellent cure as lower intensity, higher wavelength regions are able to penetrate the film more thoroughly. Examples of a D and an H bulb's spectral output can be viewed in Figs. 5 and 6, respectively. These outputs are reprinted here with permission from Fusion UV.[5] The spectral distributions are grouped into 10 nm bands, and the power output is shown in units

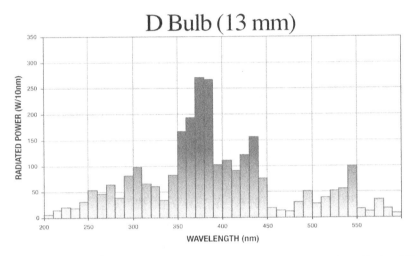

Fig. 5. Spectral output of a D bulb, reproduced with permission from Fusion UV.

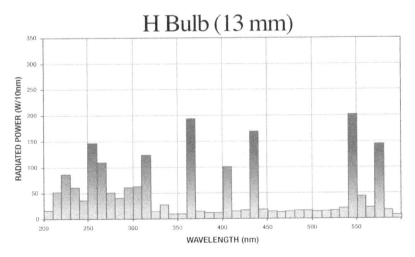

Fig. 6. Spectral output of an H bulb, reproduced with permission from Fusion UV.

of watts/10 nm. This eliminates the problem of evaluating complex emission data of medium-pressure lamps and reduces calculations of the UV spectrum to about 25 data points, which is useful for evaluation of photoinitiator or radiometer response to bulb types.

There are also less frequently used bulbs such as bulbs doped with gallium that have significant energy density in the high UV range (405–420 nm) as well as bulbs doped with indium, that emit UV in the very high UVV range (up to 450 nm). These bulbs are both proficient at achieving through cure.

Inks that can achieve a full cure using a solid state-curing unit such as an LED are still under development, as these inks must be tuned to cure not only with a low energy density, but also at a single wavelength (typically 395 nm or higher). However, if the market decides that the benefits of LED lamps, such as low energy consumption and low heat emission, are desirable enough to make the switch to LED lamps, ink makers will be forced to produce inks with similar costs and cured film properties that will cure under those conditions.

Cationic Curable Systems

Although less commonly used than free radical-based systems, UV cationic cure is still a well known technology for many formulations, including inkjet inks. This chemistry has a number of advantages such as providing excellent adhesion to metals, polyolefins, and glass. They also tend to have lower shrinkage upon curing, low odor, low toxicity are less likely to cause skin irritation, and are resistant to oxygen inhibition. Furthermore, the propagation step continues upon removal of the radiation source, unlike the polymerization process for free radical systems. However, the curing mechanism for this system is greatly inhibited by moisture since water can act as a chain transfer agent, so humidity can significantly retard the cure. Exposure to even trace amounts of acids or bases can cause the viscosity to increase and will decrease the shelf life of the formula. Also, many of the components (particularly the photoinitiators) are more expensive than their free radical counterparts.

Cycloaliphatic epoxy resins are a large part of many cationic systems, along with components such as caprolactone polyols, novolac epoxies, aliphatic epoxides, and various other polyols. Cationic systems can be initiated via UV radiation, like free radical formulations. However, the photoinitiators and curing mechanism are quite different. Aryl sulfonium or iodonium salts are typically used, with a variety of counter ions available. Irradiation via UV light results in the generation of strong acids causing a rapid ring-opening allowing the cycloaliphatic epoxies to cross link with each other and also with any hydroxyl compounds that may be present. As mentioned previously, the reaction can proceed via a "dark cure" even after the source of radiation has been removed. This is beneficial as it requires less energy density to cure completely, but it could also be problematic in the case of brief exposure to stray light that could cause the "run away" polymerization to occur in the bottle or in the ink delivery system or print head.[6]

Additives

Additives are used in many formulations to enhance certain properties, such as shelf life, flow, adhesion, and weatherability, to name a few. One property that is often manipulated for inkjet inks is the surface tension, which may be reduced by adding surfactants. Typical print heads require static surface tensions in the range of mid 20s to low 30s (dynes/cm). Fluids with static surface tensions that are too low may experience excessive nozzle faceplate wet out, which will result in loss of jetting stability. Alternately, fluids with very high surface tensions will not wet out the interior of the print heads adequately, causing a non-homogenous outflow and loss of jetting due to starvation at the nozzle as the meniscus is unable to recover rapidly enough to fire the next drop. Very high surface tensions will also affect the ability of the ink to wet out low surface energy substrates, including many plastics.

It is also important to consider the dynamic surface tension in addition to the static surface tension since the firing frequencies experienced by the fluids are very rapid (up to 40 kHz on some print heads, or 40 000 pulses per second!) and result in a dynamic environment. It has been theorized that an ideal ink would posses a high dynamic surface tension (to promote rapid meniscus recovery at high firing frequencies) and a low static surface tension (to achieve good substrate wet out).[7]

The dynamic surface tension curve can be divided into three parts: the first equilibrium state represents values at low surface ages (Region 1 in Fig. 7), a rate limiting state (Region 2 in Fig. 7), and the second equilibrium state represented by the static surface tension (Region 3 in Fig. 7).

Ideally, the equilibrium state 1 needs to be short, followed by a quick drop upon which the ink should reach the static surface tension so that the ink may wet out the substrate upon impact. Surfactants are commonly employed to lower the surface tension and increase the steepness of the decline in Region 2 on Fig. 7, so that the

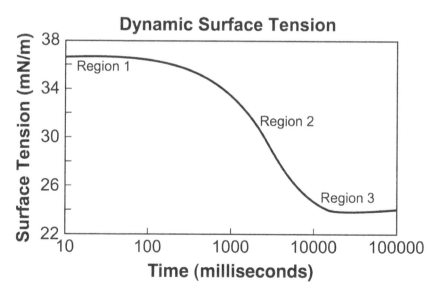

Fig. 7. Regions of interest on a dynamic surface tension plot.

surface tension of the fluid is optimized for each stage of printing. Different surfactants will affect the regions of interest differently, so it is important to evaluate several types, rather than just increasing the amount added.

Another common additive found in inkjet ink formulas are in-can stabilizers to extend the shelf life of the product. Typically, the shelf life must be at least 9 months although 12 months is preferred. Frequently, radical scavengers are used to prevent polymerization from occurring in the container. Also, it is important to leave adequate headspace in the bottle, as oxygen is also a very effective inhibitor of free radical initiated polymerization.

Other additives include materials such as hindered amine light stabilizers (HALS) and UV light absorbers (UVAs) to protect the colorants from destruction due to the environment, as well as items such as defoamers and wetting or leveling agents to help improve lay down and flow out on the media. Some of these additives such as the stabilizers and the HALS and UVAs must be added in small amounts as they can also impede curing.

STARTING POINT FORMULATIONS

Tables 2[8] and 3[9] show simple starting point formulations.

Table 2 lists only four raw materials, if the photoinitiator package is considered to be one material. However, it will have all the critical properties of a UV curable inkjet ink — low viscosity due to the monomer selection and the ability to polymerize because of the photoinitiators included. Also, the use of a pentafunctional acrylate will

Table 2. Starting point inkjet formulation.

Material	%
Dipropylene glycol diacrylate (SR508IJ)	37
Di-pentaerythritol pentaacrylate (SR399LV)	35
Pigment dispersion	20
SR1135 (blend of phosphine oxide, α-hydroxy ketone, and benzophenone derivative)	8

Table 3. Starting point inkjet formulation.

Material	%
Dipropylene glycol diacrylate (Miramer M222)	23.5
Neopentyl glycol propoxylate diacrylate (Miramer M216)	18
Laurylacrylate (Miramer M120)	17
Dipentaerythritol hexaacrylate (Miramer M600)	3
Trimethyl propane ethoxylated triacrylate (Miramer M3130)	6
Genorad 16 (stabilizer)	0.5
1-hydroxycyclohexyl phenyl ketone (Genocure LBC)	4
Liquid photoinitiator blend (Genocure LTM)	4
2-methyl-1-(4-methylthiophenyl)-2-morpolinpropan-1-one (Genocure PMP)	4
Magenta dispersion (SPF 586)	20

ensure a durable and hard cured film. Other additives and various monomers may also be added to enhance the properties required by the end user. Table 3 shows a more comprehensive formulation, complete with stabilizers, and several monomers and photoinitiator variants.

SUMMARY

Formulating UV curable inkjet inks for piezoelectric print heads requires technical knowledge as well as creativity to meet the strenuous requirements that the print markets, particularly the industrial printing arena, demand. The basic components found in most inkjet ink formulas include: monomers and oligomers, pigments and dyes, photoinitiators, and additives.

These components must be carefully selected to meet the physical property requirements for jettability — low viscosity (8–12 cps at the jetting temperature) and surface tension in the mid 20s–30s depending on the head technology chosen. It is also important to take into consideration the spectral output of the lamp being used as well as the type of media being printed upon and the end use of the printed part. As hardware advances continue to occur the demand for superior inks will also increase.

REFERENCES

1. Page A. (2006) *Developments in Radiation Curing Inks.* Pira International Ltd, UK.
2. Lindquist PA, Edison SE. (2007) Ink and equipment: The Yin and Yang of industrial printing. *Ink World* **June:** 1–5.
3. Bruice PY. (1995) Chain growth polymers. *Organic Chemistry,* pp. 1013–1019. Prentice Hall, New Jersey.
4. Balcerski J. (2006) Sartomer Corporate overview. *IMI, 5th Annual Inkjet Developers Conference,* Denver.
5. Fusion UV Systems, Inc., Gaithersburg, MD USA, copyright by Fusion UV Systems, Inc.
6. Madhusoodhanan S, Nagvekar D. (2008) Dual cure digital inks for industrial printing. *RadTech UV/EB Technology Expo and Conference 2008,* Chicago.
7. De Rossi U, Bolender O, Domanski B. (2004) Dynamic surface tension of UV-curable inkjet inks. *IS&T's NIP20: 2004 International Conference on Digital Printing Technologies,* pp. 788–792.
8. Goodrich JE. (2006) New raw materials for high temperature UV inkjet, *Sartomer Technical Paper,* www.sartomer.com.
9. Rahn EnergyCuring. (2008) SPF 598, *Rahn Starting Point Formulation,* www.rahn-group.com.

CHAPTER 10

Raw Materials for UV Curable Inks

Ian Hutchinson

INTRODUCTION

During recent years UV curable inkjet inks have gained wide acceptance as a technology that offers the printer optimum benefits. This is in part due to the unique advantages offered by UV technology. Amongst these advantages are factors such as instantaneous drying, solvent free inks, inks that adhere to a wide range of substrates and inks that are pigmented as well as possessing excellent film properties.[1] In general, these advantages are well known. However the wide variety of raw materials available for use in UV inkjet ink formulations and used to achieve these results may not be so widely known.

This chapter looks at the raw materials available to UV curable inkjet ink formulators. The main emphasis is on the performance trends of the raw materials, thus hopefully enabling formulators to produce inks that fulfill the greater performance demands required as the technology expands.

The predominant chemistry for formulating UV curable inks is acrylate-based free radical curing.[2,3] Most UV inkjet formulations are mixtures of several different acrylates chosen for their individual properties which, when combined with photoinitiators, achieve the required specification.[4] Because of the extremely low viscosity of the inkjet application, it is particularly the acrylate "monomers" which are used to achieve the end results. Therefore the greater part of this chapter will be concerned with acrylate monomers. Also discussed will be the types of acrylate oligomers that can be used to modify the formulation as well as some comments on photoinitiators. In the conclusion of this chapter there are some comments on "cationic" UV curing, which although less popular than free radical, is beginning to find niche applications.

UV CURING

The cure mechanisms of UV cured systems are well known and discussed in detail in several publications.[5,6] The use of UV curable technology in industrial applications such as wood coating and screen printing has been commercialized for over 30 years. The number of applications using UV curing technology has grown over the years such that the volumes consumed have doubled every five years. UV cured technology is very diverse. In addition to coating and printing applications it is also used in adhesives,[7] composites[8] and several electronics applications.[9] As stated earlier, all these technologies are benefiting from the key advantages of UV cured technology, which are:

- Instant drying
- Solvent free systems
- Tough resistance films
- Low energy cost application
- Adhesion to many varied substrates
- External durability
- High gloss levels.

UV technology utilizes the same building blocks in all the formulations for various applications. These building blocks are known by several different names — for example monomers, prepolymers, and oligomers. For free radical curing, they are generally based on acrylate chemistry[10] and these are blended together to achieve the final specification required by the end use application. One of the

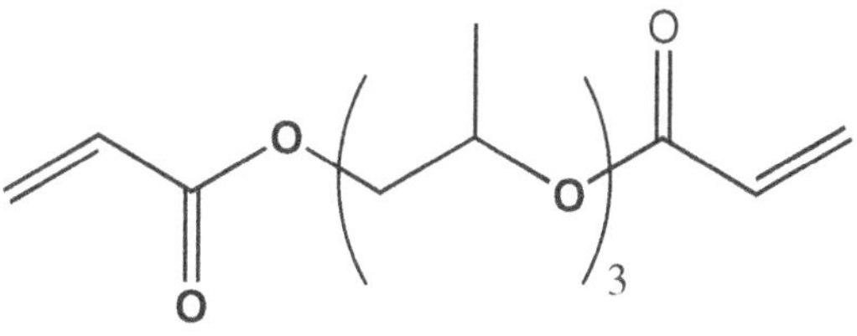

Fig. 1.   TPGDA — A typical acrylate monomer.

most commonly used acrylate monomers on the market is shown above. This is tripropyleneglycoldiacrylate (or TPGDA):

This is a typical acrylate monomer and can be used as a benchmark for comparison with other acrylate monomers. It has a viscosity of around 15 cps at 25°C which makes it quite suitable for formulating UV curable inkjet inks. There are several aspects to note in the chemical structure. It has a molecular weight of 300, it has two acrylate groups and has ether linkages. The surface tension at 25°C is 33.3 mN/m^2.[11] The commercial product will also be manufactured with stabilisers present to give the best possible balance between shelf life and cure response. What can be expected from a cured film comprised entirely of TPGDA is that it would cure at a reasonable cure speed, the film would have some physical and chemical resistance and that it would be reasonably tough. It would adhere to some substrates. In fact it is an "average" monomer. Of course TPGDA on its own would not fulfil the demands of the formulator and so other acrylates with different film properties, cure speeds and adhesion properties would need to be added to obtain all the required parameters. However, it highlights how the chemical structure is related to the overall properties of the ink and also how several different structures are required to fulfil these properties. This is discussed in more detail in the next section.

ACRYLATES — TYPES AND CHARACTERISTICS

There are probably over 800 different types of acrylate available commercially . For UV curable inks, these fall into several different categories. Of course, as was previously shown, monomers such as TPGDA have an important role. But there are also higher viscosity materials, known as oligomers available. These come in several categories — for example epoxy acrylates, polyester acrylates and urethane acrylates.[12] These materials generally have viscosities in the region of 15–100 Pas at 25°C. They are used in UV formulations to impart the highest possible film and performance properties into the formulation. However, because of their high viscosity, they can only be used at small levels in UV inkjet formulations to achieve modification of specific performance characteristics. The chemistry of the oligomers is discussed at the end of this section. Also of interest to the formulator of free radical mechanism inks are the amine modified materials such as synergists and polyesters.[13] These again have specific performance properties such as reducing oxygen inhibition effects, improving cure speed or improving wetting characteristics that can bring advantages to the formulator.

However as discussed the most important class of acrylates for inkjet applications is the monomers. Generally these range from monoacrylate to hexaacrylate functionality and in viscosity from 5 cps up to around 150 cps. The key characteristics of each monomer required by the formulator for selection can be listed thus:

- Acrylate functionality
- Chemical structure
- Molecular weight
- Viscosity
- Surface tension
- Shrinkage on cure
- Pigment wetting properties
- Odour.

In addition to this, there are the properties required in cured ink, which are directly related to the characteristics of each monomer:

- Reactivity
- Adhesion
- Physical resistance of the cured film

- Chemical resistance of the cured film
- Flexibility
- Gloss.

Acrylates tend to react differently based on acrylate functionality. A typical comparison of the reactivity of mono-, di- and tri-acrylates is shown in Fig. 2.

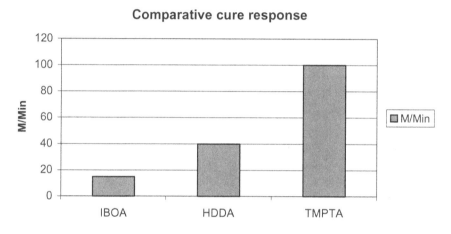

Fig. 2. Typical comparison of cure response (mono-, di- and tri-acrylate).[6]

As would be expected, these properties vary widely as the monomer functionality increased. The chart below indicates the general trends that can be expected as acrylate functionality increases.

Table 1. General overview of comparative acrylate properties.

	Monoacrylate	Diacrylate	Triacrylate	Tetraacrylate
Cure response	Slow	Moderate	Fast	Fastest
Flexibility	Very flexible	Flexible	Rigid	Brittle
Film hardness	Softest	Moderate	Hard	Hardest
Solvent resistance	Generally low	Moderate	Good	Best
Adhesion	Good	Good	Moderate	Poor

As another example of the trend between acrylate functionality, films created from several monomers have been cured and their Tg measured.

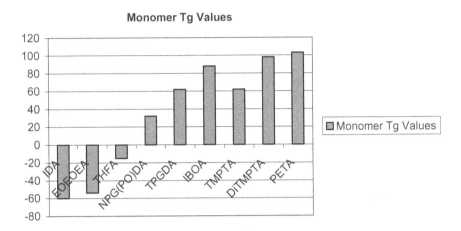

Fig. 3.   Monomer cured film Tg values.[6]

Of course these trends in performance as the acrylate functionality increases are only general guidelines. Within each category there are other factors which determine the actual performance of a monomer. These are derived from the chemistry of each individual monomer. Within each monomer category there can be several variations.[3]

Factors to be taken into consideration are:

a. *Effects of the monomer being linear or cyclic.*
   Generally cyclic monomers tend to have tougher film properties and are less flexible. Linear monomers (particularly monoacrylates) are more flexible.
b. *Effects of the cyclic monomer being aromatic or aliphatic.*
   Aromatic monomers have very good chemical resistance but do not have the external durability characteristics of aliphatic monomers. Aliphatic monomers will have better non-yellowing properties.

c. *Whether the monomer is alkane or ether.*
Alkane based monomers will be more hydrophobic and have better external durability characteristics. They will tend to be non-yellowing and promote adhesion. Ether based monomers will give improved cure resistance, chemical resistance but will be quite flexible.

d. *Effect of monomer alkoxylation (ethoxylated or propoxylated).*
Generally alkoxylated monomers will be more flexible and have higher viscosity than the non-alkoxylated countertypes. They will also have lower skin irritancy and ethoxylated monomers will be faster curing whereas propoxylated monomers give more rigid films but benefit from having improved pigment wetting properties.

As the degree of ethoxylation increases, the monomer will become more hydrophilic, have better water solubility and surface cure. The effect of increasing propoxylation is to improve pigment wetting and to decrease surface cure speed and water solubility.[14]

All of these parameters can be taken into consideration when assessing the suitability of monomers for specific ink formulations. There is a wide range of monomers available commercially for use in inkjet formulations and these are outlined below in more detail.

Linear Monoacrylate Monomers

The general trend with these materials is high flexibility and adhesion but slow cure response.

2(2-Ethoxyethoxy)ethyl acrylate

- Viscosity 5 mPas at 25°C
- Highly flexible monomer with good adhesion to plastic substrates.

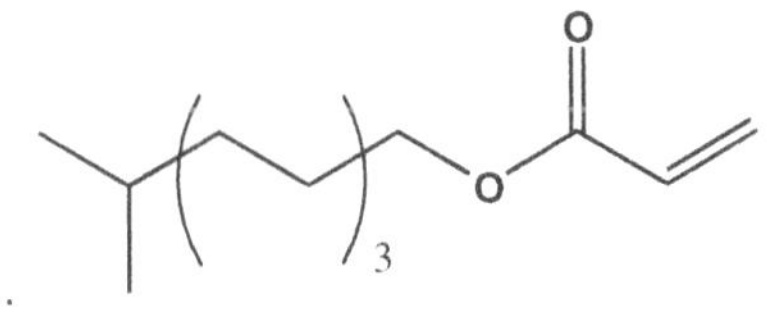

Isodecyl acrylate

- Viscosity 10 mPas at 25°C
- Very flexible monomer with low surface tension.

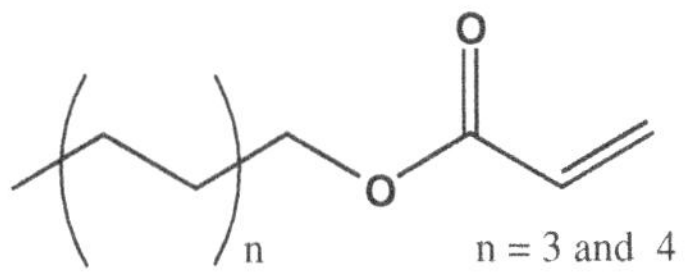

Octyl/decyl acrylate

- Viscosity 10 mPas at 25°C
- Long aliphatic backbone monomer with good wetting properties.

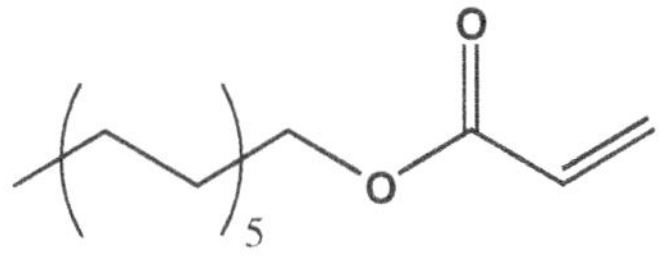

Lauryl acrylate

- Viscosity 6 mPas at 25°C
- Very soft flexibilizing monomer.

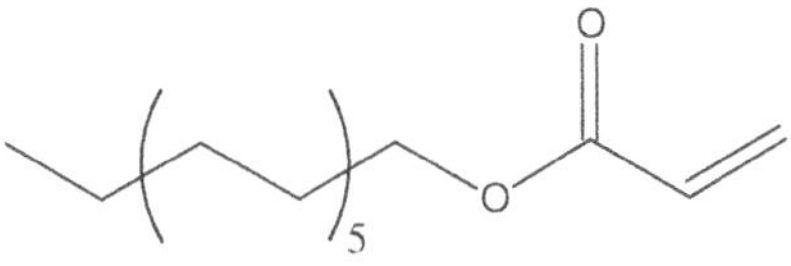

Tridecyl acrylate

- Viscosity 8 mPas at 25°C
- Low odour monomer with high flexibility.

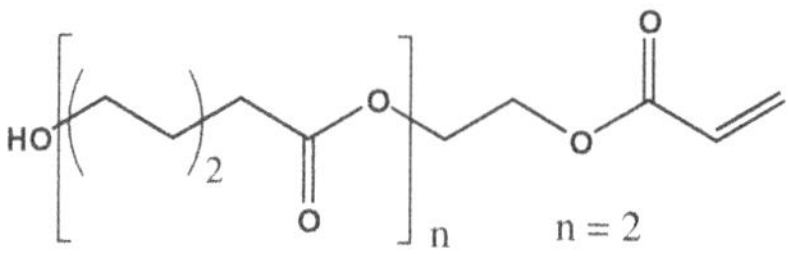

Caprolactone acrylate

- Viscosity 75 mPas at 25°C
- Low odour monomer with good film properties.

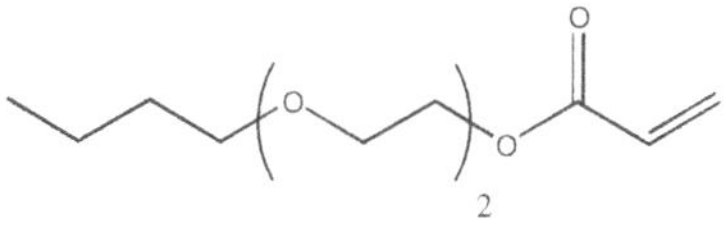

Diethylene glycol butyl ether acrylate

- Viscosity 5 mPas
- Low viscosity monomer with good adhesion and film properties.

Cyclic Monofunctional Monomers

These monomers generally give better film properties than the linear monofunctional monomers. However in some cases they have quite distinct odours.

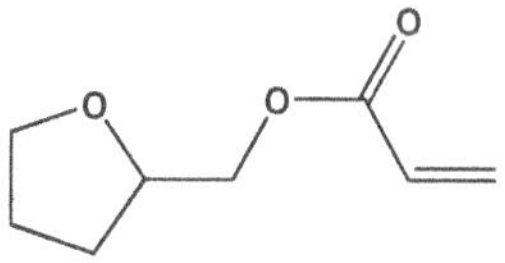

Tetrahydrofurfuryl acrylate

- Viscosity 5 mPas at 25°C
- A very aggressive monomer with good adhesion onto many plastics.

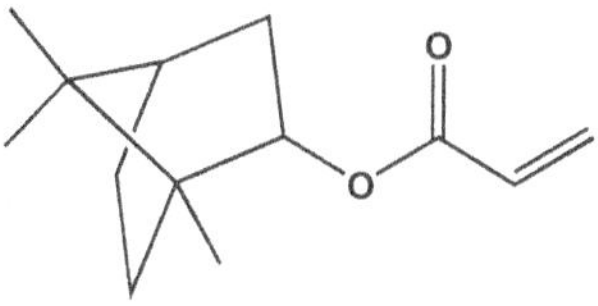

Isobornyl acrylate

- Viscosity 10 mPas at 25°C
- A monomer which gives very tough films and good adhesion.

Cyclic trimethylolpropane formal acrylate

- Viscosity 12 mPas at 25°C
- A low odour monomer with good cure response and adhesion characteristics.

Isophoryl acrylate

- Viscosity 6 mPas at 25°C
- Low shrinkage monomer with high impact strength.

Aromatic Monofunctional Monomers

These monomers give tough films with excellent solvent resistance. However, the adhesion may not be as good as the aliphatic cyclic monomers.

2-Phenoxyethyl acrylate (2-PEA)

- Viscosity 10 mPas at 25°C
- Tough film forming monomer with good adhesion properties.

Ethoxylated (4) phenol acrylate

- Viscosity 35 mPas at 25°C
- An ethoxylated version of 2-PEA with improved flexibility, odour and skin irritancy.

Ethoxylated (4) nonyl phenol acrylate

- Viscosity 90 mPas at 25°C
- Highly flexible, low odour monomer.

Difunctional Acrylate Monomers

These are the general standard monomers for inkjet applications. They possess the best compromise of viscosity, cure speed and film properties required for the formulation. Several different types are available as shown.

Hexanediol diacrylate

- Viscosity 7 mPas at 25°C
- Standard difunctional acrylate with excellent adhesion properties and low viscosity.

Tricyclodecane dimethanol diacrylate

- Viscosity 120 mPas at 25°C
- Highly versatile monomer with good adhesion and film hardness.

Dioxane glycol diacrylate

- Viscosity 250 mPas at 25°C
- A fast curing monomer giving tough films and good adhesion.

Dipropylene glycol diacrylate (DPGDA)

- Viscosity 10 mPas at 25°C
- Multipurpose difunctional monomer.

Tripropylene glycol diacrylate

- Viscosity 15 mPas at 25°C
- More flexible than DPGDA but less reactive.

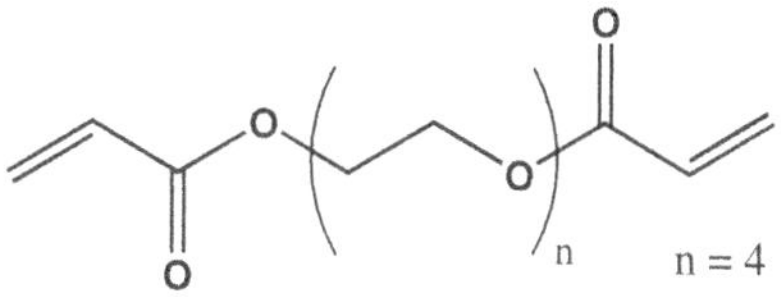

n = 4

Polyethylene glycol (200) diacrylate

- Viscosity 25 mPas at 25°C
- Very flexible diacrylate monomer.

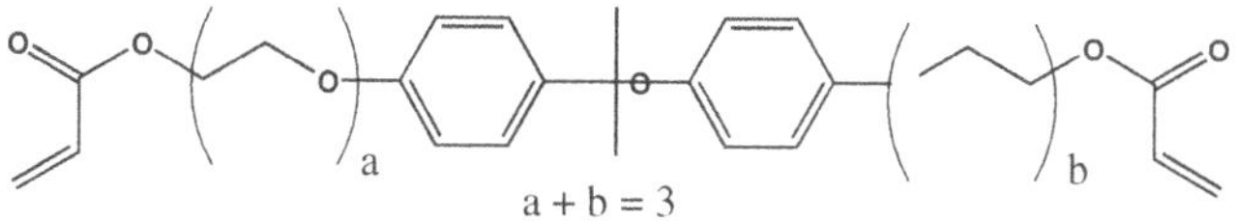

a + b = 3

Ethoxylated bisphenol A diacrylate

- Viscosity 1500 mPas at 25°C
- Very tough, fast curing monomer used to modify film performance.

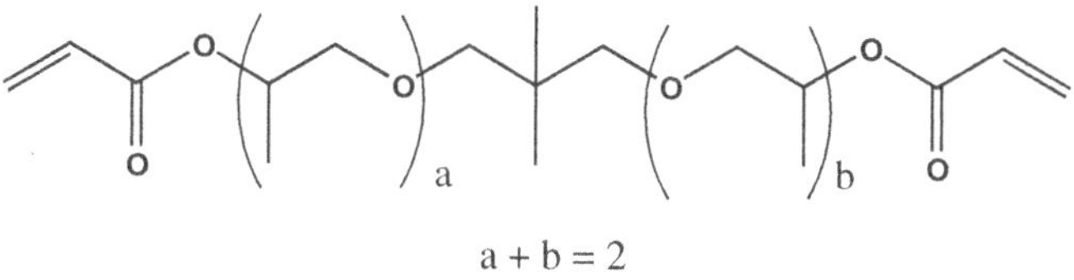

a + b = 2

Propoxylated (2) neopentyl glycol diacrylate

- Viscosity 18 mPas at 25°C
- The propoxylation gives a monomer with excellent pigment wetting and adhesion properties.

Trifunctional Acrylate Monomers

With these materials, increasing cure speed is observed but also with increasing viscosity.

Trimethylolpropane triacrylate

- Viscosity 110 mPas at 25°C
- Fast reacting, tough monomer.

$$a + b + c = 3$$

Propoxylated (3) trimethylolpropane triacrylate

- Viscosity 100 mPas
- Version of TMPTA which has better pigment wetting and flexibility.

$$a + b + c = 3$$

Ethoxylated (3) trimethylolpropane triacrylate

- Viscosity 70 mPas at 25°C
- Version of TMPTA with good cure response and flexibility
- Higher ethoxylates up to 15(EO) which exhibit even more flexibility and lower odour are also available.

a + b + c = 3.5

Propoxylated glycerol triacrylate

- Viscosity 90 mPas at 25°C
- Triacrylate with good pigment wetting and cure response.

Tris(2-hydroxylethyl)isocyanurate triacrylate

- Solid at 25°C but soluble in many monomers
- Ideal additive to impart hardness into a cured film.

Tetra and Higher Functional Monomers

The use of these materials is generally to increase cure response and film hardness. However, they generally will have an adverse impact on adhesion as well.

Dipentaerythritol penta/hexa acrylate

- Viscosity 8000 cps
- Very fast curing monomer with excellent film hardness.

Alkoxylated pentaerythritol tetraacrylate

- Viscosity 150 mPas at 25°C
- Fast curing flexible monomer.

Di(trimethylol)propane tetraacrylate

- Viscosity 600 mPas at 25°C
- Flexible fast curing monomer.

ACRYLATE OLIGOMERS

In general, the acrylates classified as "oligomers" are too high in viscosity for extensive use in the very low viscosity formulations required by UV curable inkjet inks. However, acrylate oligomers do possess several properties that can be very beneficial, even when used at additive levels.

There are three main types of oligomers, epoxy acrylate, urethane acrylate and polyester acrylate. The basic trends in properties are demonstrated in Table 2.

There are also some other benefits to be gained from using some oligomers in a formulation. For example, the very high degree of pigment dispersion required to fulfil the ink specification may be better achieved by the use of a specifically designed oligomers (normally a polyester acrylate). The higher molecular weight of some of the urethane acrylates also lends itself to improving flexibility of some films.

Table 2. Comparison of characteristic properties of oligomer types.[15]

	Epoxy Acrylate	Urethane Acrylate	Polyester Acrylate
Viscosity	High (80–100 Pas)	Variable (1–100 Pas)	Variable (1–80 Pas)
Acrylate functionality	2	2–6	2–6
Molecular weight	500–800	500–5000	500–2500
Cure speed	High	Variable	Medium
Tensile strength	High	High*	Medium
Flexibility	Low	High*	Low
Adhesion	Poor	Good	Average
Chemical resistance	Very Good	Good	Average
Hardness	High	High*	Moderate
Yellowing	High	Low*	Low
Brittleness	High	Low*	High

* For optimal cases only. Urethane acrylates possess a wide range of properties depending on the nature of the individual material.

Epoxy Acrylates

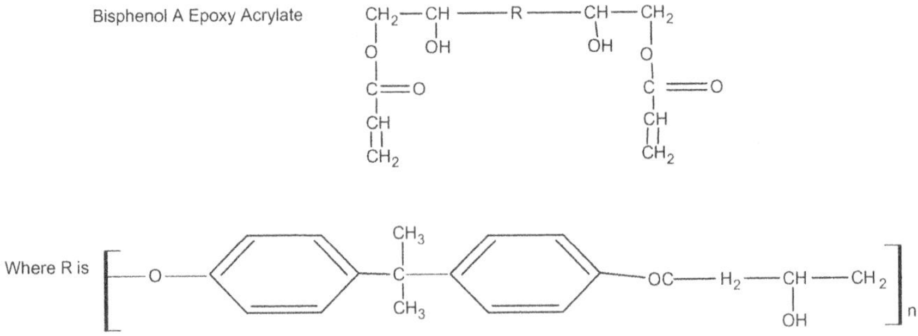

Bisphenol A Epoxy Acrylate

Where R is

Epoxy acrylates are formed by the addition of acrylic acid to bisphenol A diglycidyl ether thus producing the above chemical structure.[16] The aromatic ring structure is responsible for many of the properties of epoxy acrylates.

There are variations to the above structure such as fatty acid modification but these tend to increase the viscosity significantly thus limiting use in inkjet ink formulations.

Urethane Acrylates

Urethane acrylates are produced by the reaction of up to three raw materials — an isocyanate, a hydroxyl functional capping agent such as hydroxyl ethyl acrylate and a polyester or polyether polyol. The scope of raw materials available allows urethane acrylates to be "designed" for a specific end use. This means urethane acrylates can have a very wide array of film properties ranging from very flexible to very tough. However, the viscosity of urethane acrylates is normally quite high and once again its usage may be limited in inkjet formulations.

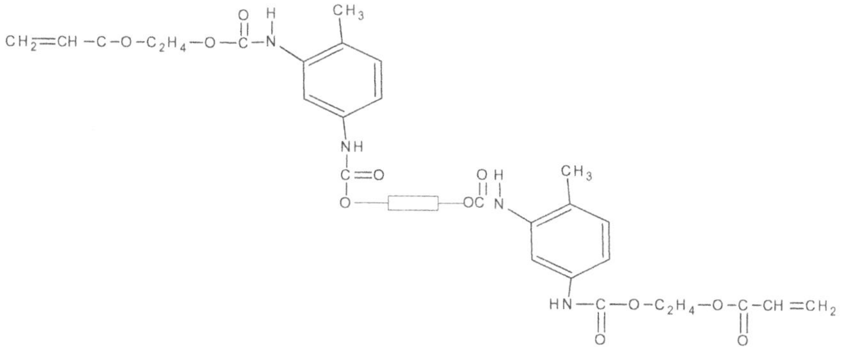

A typical urethane acrylate.

Polyester Acrylates

Polyester acrylates are normally formed by the reaction of polyester or polyether polyols reacted with acrylic acid. As with urethane acrylates, this gives the chemical structure of polyester acrylates great versatility. They tend to have lower viscosity than urethane acrylates but with inferior film properties. However, the low viscosity of some polyester acrylates does allow them some uses in UV curable inkjet ink formulations.

Some speciality polyester acrylates are designed for improved pigment wetting. Some are designed for low viscosity and flexibility. A typical polyester acrylate is shown below.

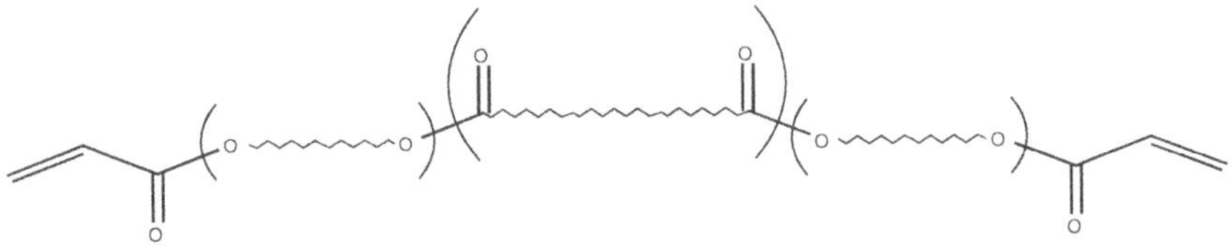

AMINE MODIFIED POLYESTERS AND SYNERGISTS

One of the means used to modify monomers and polyester acrylates further is to amine modify them. This is normally done using a Michael addition reaction between the acrylate and an amine.[17] The benefits of amine modification are normally seen with increased cure speed since this will tend to overcome the effects of oxygen inhibition in the cure process. Amine oligomers and synergists can also be low viscosity and give improved flexibility to a film. One disadvantage of these materials maybe that perhaps they do not possess the stability required of UV inkjet ink formulations. They are also very prone to yellowing and are unstable with some forms of adhesion promoters.

As an indication of the benefits to surface cure by adding amine modified synergists, the figure below compares the reactivity of a standard formulation. The comparison shows the difference between no amine, liquid amine (n-methyldiethanolamine) and two types of synergist ("Syn" 1 and 2) which have amine values of 150 and 200 mg KOH/g.

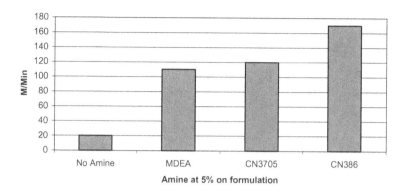

Fig. 4.   Effect of amine synergist on cure speed.[18]

PHOTOINITIATORS FOR FREE RADICAL SYSTEMS

The selection for the photoinitiators is one of the key elements of formulation. Generally blends of different photoinitiators have to be used as a "package" in order to achieve optimal cure response. It is not uncommon to use a blend of four different photoinitiators to achieve cure. The photoinitiator package has to absorb the UV light such that it allows surface cure of the ink film and through cure of the ink film at the required speed of the printer to give optimum properties. It also has to effect the cure of pigmented systems and overcome any effects caused by oxygen inhibition.

Therefore, there are many different types of photoinitiators available for the formulator. Each has specific properties. Some of the more common ones used in UV inkjet formulations are listed here.

Photoinitiator	Property
Benzyl Dimethyl Ketal	Used through cure of thick coatings.
(Irgacure651, BDK)	Tends to yellow.
2-Hydroxy-methyl-1-phenyl propane (Darocure 1173)	Fast curing for use in pigmented systems.
Hydroxycyclohexylphenylketone	Surface cure.

(Continued)

(*Continued*)

Photoinitiator	Property
(Irgacure 184, HCPK) Irgacure 907	Non-yellowing. Effective with pigmented systems. Requires co-initiation.
Irgacure 369	For pigmented systems.
Monoacyl phosphine oxide (Lucerin TPO)	For pigmented systems with TiO_2.
Esacure KIP150	Polymeric photoinitiator. Improved film properties.

TYPICAL FORMULATION

The formulation of UV curable inkjet inks is built up from all the raw materials described in previous sections. In order to achieve the cure speed, adhesion and other final film properties, a combination of raw materials are generally required. It is also the case that there are many patents covering this area and so formulators should respect this.

A typical formulation could potentially look like this:

Material	%	
Pigment	3	
Wetting additive	1	For pigment dispersion.
Low viscosity oligomer	10	For cured film resistance and flexibility.
Triacrylate monomer	12	For cure speed — e.g., TMP(EO)TA.
Diacrylate monomer	22	For cure speed and adhesion — e.g., TPGDA.
Monoacrylate monomer	38	For viscosity and adhesion — e.g., CTFA or IBOA.
Additives	2	For deaeration, surface wetting.
Photoinitiator blend	12	

CATIONIC SYSTEMS

Cationic cured systems have traditionally been aimed at niche applications or where ink based on free radical acrylate curing is not suitable. The raw materials used in cationic systems are generally different than those used in free radical cured systems.[19]

The advantages of cationic systems are significant when compared to free radical acrylate based inks:

- Not affected by oxygen inhibition.
- Excellent adhesion to a wide variety of substrates.
- Physical film properties that exceed those of acrylate systems.
- Very low viscosity of vinyl ether diluents.

However there are disadvantages:

- Curing is easily fouled by humidity and amines.
- Pigment choice is limited.
- Extra cost may be incurred by using post-cure thermal treatment.

But the key disadvantage is the narrow range of raw materials available to the formulator, thus restricting the flexibility allowed by the system. A typical formulation may contain vinyl ether diluents, a cycloaliphatic epoxide as the film forming material and photoinitiator.

With respect to the photoinitiator and cycloaliphatic epoxide, there are fewer than ten products on the market. The choice is higher for vinyl ether diluents.

REFERENCES

1. Edison SE. (2006) UV-curable inkjet inks revolutionize industrial printing. *RadTech Report* **20**(6): 28, 30–33.
2. Visconti M, Cattaneo M. (2000) A highly efficient photoinitiator for water-borne UV-curable systems. *Prog Org Coatings* **40**: 243–251.
3. Dietliker K, Oldring PTK. (1991) Free radical polymerization. In Oldring PKT (ed.), *Chemistry and Technology of UV and EB Formulation for*

Coatings, Inks and Paints. Vol. 3, pp. 75–77. Selective Industrial Training Associates Technologies Ltd., London.

4. Fuchs A, Villeneuve S, Richert M. (2004) European Patent WO2004092287 (A1).

5. Selli E, Bellobono IR. (1993) Photopolymerisation of multifunctional monomers: kinetics aspects. In Fouassier JP, Rabek JP (eds.), *Radiation Curing in Polymer Science and Technology: Polymerization Mechanism*, pp. 1–28, Springer.

6. Sartomer Europe Product literature.

7. Czech Z. (2007) Synthesis and cross-linking of acrylic PSA systems. *J Adhes Sci Tech* **21**(7): 625–635.

8. Shukla V, Bajpai M, Singh D, Shukla R. (2004) Review of basic chemistry of UV-curing technology. *Pigment & Resin Technology* **33**(5): 272–279.

9. Willard K. (2000) UV/EB curing for automotive coatings. *RadTech report* **14**(6): 22–27.

10. Arifuki M, Watanabe I, Goto Y, Kobayashi K, Nakazawa T. (2005) Material for connecting circuits, and connected structure of the circuit component using this material. *Japanese Kokai Tokkyo Koho*, 23 pp.

11. Sartomer product bulletin: SR306F (Trypropylene glycol Diacrylate).

12. Rad-Solutions, LLC US Product literature.

13. Kranig W, Blum R. (2001) US Patent 6177144.

14. Asai T, Swanson BL, Tomko SE, Fernyhough A, Fryars M. (2000) US Patent 6103317.

15. Allen NS, Johnson AM, Oldring PKT, Salim MS. (1991) Prepolymers and reactive diluents for UV and EB curable formulations. In Oldring PKT (ed.), *Chemistry and Technology of UV and EB Formulation for Coatings, Inks and Paints*, Vol. 2, pp. 228–236. Selective Industrial Training Associates Technologies Ltd., London.

16. Michailov YM, Ganina LV, Smirnov VS. (2002) Phase equilibrium in biphase polymer systems based on Diglycidyl Ether of Bisphenol A. In Rozenberg BA, Sigalov GM (eds.), *Heterophase Network Polymers: Synthesis, Characterization, and Properties*, pp. 33–42, CRC Press.

17. Chan JW, Zhou H, Wei H, Hoyle C. (2008) Mechanism of the primary and secondary amine catalyzed thio-Michael addition reaction with (meth)acrylates. *Abstracts of Papers, 235th ACS National Meeting, New Orleans, LA, United States, April 6–10*, **2008**: 168 pp.

18. Oliver J. (2007) The use of amine synergists for low odor and low migration applications. *Radtech Conference.*

19. Crivello JV, Dietliker K. (1991) Photoinitiators for cationic polymerization. In Oldring PKT (ed.), *Chemistry and Technology of UV and EB Formulation for Coatings, Inks and Paints.* Vol. 3, pp. 329–373. Selective Industrial Training Associates Technologies Ltd., London.

CHAPTER 11

Unique Inkjet Ink Systems

Matti Ben-Moshe
DIP-Tech Ltd.

Shlomo Magdassi
The Hebrew University of Jerusalem

INTRODUCTION

Inkjet printing is a digital method for direct printing of computer data onto a substrate, such as paper, transparency, polymers, or ceramic substrate. One of the greatest challenges in the development of inkjet printing technologies is the ink formulation. A typical inkjet ink for the graphic arts is usually comprised of a variety of components including polymers, dyes or pigments, dispersants, and other additives in the solvent system. This solvent system, which incorporates a variety of components and solids, requires careful selection so that the final ink will be kinetically or, most preferably, thermodynamically stable for long periods of time.

This chapter describes novel inkjet inks based on a variety of vehicles, and demonstrates several optical applications utilized by inkjet inks. It aims to provide a general description of inks which are based on unique components and structures, mainly micellar systems, polyelectrolyte complexes, microemulsions, miniemulsions, emulsions, liquid crystals, and interesting phase transition inks. Utilization of self-assembly of colloidal particles and hybrid inks for fabricating optical devices will be described.

MICELLAR AND MICROEMULSION INKJET INKS

A micelle is a colloidal aggregate of amphiphilic molecules (50–100 molecules per micelle) which forms at a specific concentration termed the critical micelle concentration. As illustrated in Fig. 1, in polar media such as water, the hydrophobic part of the amphiphilic molecule tends to locate away from the polar phase while the polar groups of the molecule tend to locate in the water phase, forming the micelle aggregate. Micellar systems are able to solubilize both hydrophobic and hydrophilic compounds.

Surfactants are usually added to inkjet inks to adjust the surface tension of the ink. Too high surface tension may result in improper jetting through the print head and insufficient wetting of the printed substrate, resulting in inhomogeneous print quality and long drying time. Surface tension which is too low will cause dripping through the print head nozzles and orifice plate wetting. Low surface tension may also cause the ink to smudge and feathering might occur on the printed substrate.

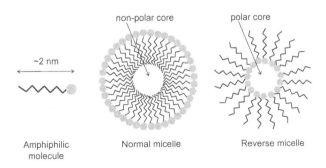

Fig. 1. Schematic form of an amphiphile and scheme of normal and reverse micelles.

Color bleed alleviation was reported[1] when employing various types of ionic or non-ionic amphiphiles in the inkjet ink composition. Color bleed is the "invasion" of one color into another on the printed surface. The amount of amphiphile is given relative to the critical micelle concentration (cmc). Above the cmc, micelles that solubilize the dye molecules form and thus control the color bleeding phenomenon. Below the cmc, free amphiphiles are present with no micelle formation, and thus they have no control over the color bleeding.

Winnik *et al.*[2] described an aqueous heterophase inkjet ink composition containing a dye covalently attached to chemical groups consisting of poly(ethylene glycols) or poly(ethylene imines), which are complexed with a heteropolyanion selected from the group consisting of phosphotungstic acid, phosphomolybdic acid, silico tungstic acid, and dichromic acid or their salts. These ink compositions are reported to have superior optical print densities, and have excellent waterfastness characteristics due to the complexed dyes in the inks, enabling the dye to be chemically protected. When these compositions impinge on the paper during inkjet printing, they precipitate immediately onto the paper fibers. Accordingly, such particles separate from the colorless solvent rather than penetrating undesirably into the paper.

Another approach which was presented by Moffatt *et al.*[3] described aqueous inkjet inks comprised of water-soluble amphiphilic dye having a chromophore and a hydrophobic tail attached. This amphiphilic dye is present at a concentration greater than the critical micelle concentration (cmc) for the dye such that micelles are formed that incorporate the dye residue. These inks are described as having a pKa value greater than the pKa of the surfactant, otherwise the amphiphilic dye loses its waterfastness. If the pH is too low, the amphiphilic dye is insoluble in the ink vehicle.

Liposomes were also utilized for preparing inkjet inks. Liposomes are aqueous compartments[4] enclosed by lipid bilayer membranes which form spontaneously when amphiphilic lipid molecules are dispersed in water (see Fig. 2). Liposomes are also known as lipid vesicles.

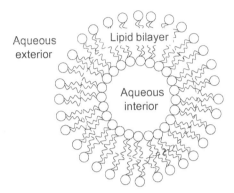

Aqueous exterior

Lipid bilayer

Aqueous interior

Fig. 2. Schematic illustration of a single bilayer membrane liposome in an aqueous phase.

Breton *et al.*[5] developed an inkjet ink composition containing liposomes. The aqueous ink contains a dye and a vesicle-forming lipid. The vesicles have membranes of lipid molecules which are covalently bonded to other lipid molecules in the membrane by polymerization. The ink was prepared by mixing water and a vesicle-forming lipid having functional groups capable of reacting with a reactive dye, causing vesicles of the lipid to form in the aqueous solution. The admixing of the aqueous solution with a reactive dye at alkaline pH causes the dye to be covalently bonded to the lipid vesicles. The vesicles in the ink have an average particle diameter of less than about 200 nanometers. Breton *et al.* reported that for black inks, particle diameters of up to about 500 nanometers may be suitable. Particle size of the vesicles can be adjusted by several methods, such as by the choice of starting materials and the substituents, chain length of the hydrophobic portion, and by filtration through a controlled pore size. The ink ingredients may affect the size and stability of the liposomes. For example, a vesicle-forming lipid may form liposomes of about 50 nanometers in diameter, but when other ingredients, such as organic co-solvents and humectants commonly employed in thermal inkjet inks are added to the composition, the liposomes may swell to diameters of about 100 nanometers because the less polar organic material may have a tendency to push apart the "tails" in the interior of the liposome bilayer membrane, thereby thickening the membrane and possibly rendering it less stable.

Microemulsion-based inkjet inks are one of the many inkjet ink formulations which have been suggested in the past years. Microemulsion-based inks may provide a fast dry, bleed control, and waterfast advantages to the formulation and, above all, a thermodynamically stable ink, which forms spontaneously.

Microemulsions are thermodynamically stable, clear isotropic liquid mixtures of water, oil, and surfactant, usually comprising a co-surfactant. The aqueous or oil phase of the microemulsion may contain dissolved ingredients such as salts, dyes, or reactive materials. Unlike emulsions (which are presented below), microemulsions form spontaneously upon simple mixing of the components and do not require the high shear conditions generally used in the formation of ordinary emulsions. The three basic types of microemulsions are direct microemulsions (oil-in-water – o/w), reversed (water-in-oil – w/o), and bicontinuous. In microemulsions, the immiscible water and oil phases are present in the system together with a surfactant. The surfactant molecules form an interfacial layer between the oil and water; the hydrophobic tails of the surfactant molecules are in the oil phase and the hydrophilic head groups in the aqueous phase. This arrangement induces self-assembled structures of different types, from spherical (inverted) and cylindrical microemulsions to lamellar phases and bicontinuous microemulsions. The typical size of microemulsion droplets is less than 30 nm, making them very attractive for inkjet applications.

Several microemulsion inkjet inks have been described in the literature. An inkjet phase transition ink in the form of a microemulsion[6–8] consists of an organic vehicle phase having a colorant dispersed therein, where the vehicle phase is preferably liquid while jetting at temperatures above 70°C and solid upon keeping the substrate at room temperature (22–25°C). This formulation undergoes a phase transition from a microemulsion phase to a lamellar phase upon heating, which allows build up of several layers of inks on the surface of the paper. In a similar concept[9] for phase transition, an ink comprised of an aqueous phase, an oil phase, an oil-soluble dye, and a surfactant is present as a liquid crystalline gel phase at an initial temperature and a liquid microemulsion at a second, higher temperature. The idea in this approach is to solubilize water-insoluble dyes in the droplets, and thereby producing waterfast images.

In another microemulsion system,[10-12] the colorant is incorporated into the ink as an aqueous pigment dispersion-based inkjet ink composition by formulating the ink to comprise at least one aqueous pigment dispersion and a microemulsion with at least one water-insoluble organic compound, one hydrotropic amphiphile, and water. This ink system was reported to improve waterfastness and bleed control,[13-15] providing a fast drying ink.

In a different microemulsion system,[16,17] a thermal inkjet ink is described, containing a vehicle and a colorant or water-insoluble colorant. The vehicle is comprised of a microemulsion containing water-insoluble organic oil, organic co-solvent, water and, optionally, amphiphile and high molecular weight colloid. The colorant is a water-insoluble chromophore that has been chemically modified to be water-soluble by the addition of functional groups that impart water solubility. The inks described demonstrate high edge acuity, high optical density, fast drying time, reduced bleed, improved halo characteristics, high water fastness, and high smear fastness. These patents suggest, for the first time, the incorporation of an oil-soluble dye into a microemulsion which contained solvents, surfactants, co-surfactants, and water.

We have developed[18,19] an oil-in-water and bicontinuous microemulsion inkjet ink composition comprising a solubilized hydrophobic dye which forms nanoparticles ("pigment-like") upon application on a substrate surface. The concept was demonstrated for direct patterning of water-insoluble organic molecules in the form of nanoparticles. The method is based on formation of thermodynamically stable oil-in-water microemulsions, in which volatile "oil" contains the dissolved organic molecules. As schematically illustrated in Fig. 3, the microemulsion droplets are converted into organic nanoparticles upon impact with the substrate surface due to evaporation of the volatile solvent.

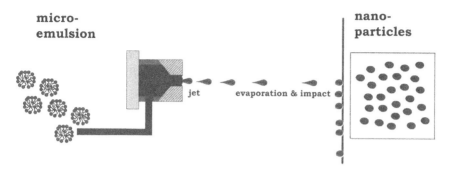

Fig. 3. Schematic representation of inkjet printing of microemulsion inks forming nanoparticles upon drying on a surface.

By this approach, it is possible to use a variety of colorants and other functional materials which are insoluble in water, which otherwise would have to be used as pigments. Obviously, preparation of nanometric pigment is not simple, and requires intensive research and development activities, which should be focused on each individual colorant (or functional molecule). All that is required in this approach is to dissolve the functional molecule in a volatile organic solvent, and find the proper composition for obtaining the microemulsion. This ink will be stable forever (at a given temperature range), and will not clog the print head nozzles, since it will behave as a dye as long as it is not dried.

EMULSION AND MINIEMULSION INKJET INKS

An emulsion is a mixture of two immiscible solvents. One substance, designated the dispersed phase, is dispersed in the other (the continuous phase). Emulsions have a cloudy appearance because of the large droplet size of the dispersed phase and the multiple interfaces that scatter the light which passes through the emulsion. Emulsions are kinetically stable and do not form spontaneously. The preparation of emulsions is associated with energy input through stirring at high shear or by spray processes. The droplets in conventional emulsions, such as milk and cosmetic products, are in the micron size range. A special type of emulsion is the miniemulsion, which is prepared by shearing a mixture of two immiscible solvents, a surfactant and, optionally, a co-surfactant. The shearing is performed with a high-pressure homogenizer or by ultra-sonication of the mixture. Stable droplets are then obtained, which have a typical sub-micron size, ranging from 50 to 500 nm.

The use of miniemulsions for inkjet printing has been described by several authors. Wong *et al.*[20] described the use of an emulsion or suspension of an organic phase in a water phase, so that the organic phase includes oil and wax. Caputo[21] described a stable oil-in-water ink emulsion containing dissolved solvent dyes. The systems contain an organic solvent of citrus origin, and a polymeric binder. Spinelli[22] described emulsion-based ink compositions of an aqueous continuous phase; a discontinuous, non-aqueous phase comprising a pigment, a dispersant, and a non-aqueous carrier; and an emulsifier that exhibits excellent resistance to water and smear, and dramatically improves bleed control in printed patterns. The invented inks are stable, have low viscosity, exhibit excellent print quality, provide excellent smear resistance after drying, and reduced bleeding. These emulsion-based ink compositions may be used with a variety of inkjet printers such as continuous, piezoelectric, drop-on-demand, and thermal or bubble jet drop-on-demand, and are particularly adapted for use in thermal inkjet printers.

Kabalnov[23] described water-based inkjet ink compositions that are miniemulsions, i.e., an aqueous vehicle having emulsified oil particles with dissolved dye molecules, where the oil droplets have a diameter of less than 1 μm. In his patent, Kabalnov mentioned the advantages of miniemulsions in comparison to microemulsions, namely the surfactant nature and concentration which allow better penetration control to the printed papers, and the dye load in miniemulsions can also be increased compared to microemulsions at acceptable viscosity limits. According to this invention, the aqueous inkjet ink composition is comprised of an oil-soluble dye, a solvent, and an aqueous vehicle wherein particles of the oil-soluble dye are dissolved in low-polarity oil particles having a particle size of less than 1 μm, the particles forming miniemulsions in water.

The method of preparing the miniemulsion inkjet ink is by the following steps:

a. dissolving an oil-soluble dye in a oil/solvent system to form a dye-containing oil;
b. adding the dye-containing oil to an aqueous vehicle;
c. mixing the dye-containing oil into the aqueous vehicle to form a mixture;
d. emulsifying the mixture to form an oil-in-water emulsion having dye-containing oil particles no larger than 1 μm in diameter.

As described above, the development of emulsion-based inks has evolved but major issues still limit their development. The microemulsion print quality is usually limited due to the ink penetration into the paper. As for the miniemulsions, better particle size control and improved emulsion stability are needed. In addition, more environmentally friendly solvents that have less smell, lower toxicity, and higher solubilizing power of the colorant need to be developed.

POLYELECTROLYTE INKS

An interesting approach for printing complex structures was suggested by Lewis *et al.*,[24] based on ink composed of oppositely charged polyelectrolytes which rapidly coagulate to yield self-supporting filaments upon deposition into the coagulation reservoir. Printing of inkjet inks comprised of concentrated electrolyte solutions with high viscosity imposes many jetting problems. It has been suggested to design inks which are composed of non-stoichiometric mixtures of polyanion and polycations. By regulating the ratio of anionic to cationic groups and combining these species under solution conditions which promote polyelectrolyte exchange reactions, the researchers obtained fluids possessing the required viscosities needed to flow through microcapillary nozzles of varying diameters. Upon deposition into an alcohol/water reservoir, the polyelectrolyte rapidly coagulates, leading to elastic ink filaments which can promote shape retention while maintaining sufficient flexibility for continuous flow and adherence to substrates. Using these fluids, periodic 3D scaffolds were created within water-alcohol solutions.

INKJET PRINTING OF OPTICAL DEVICES

Inkjet Printing of Colloidal Photonic Crystals

Photonic crystals are periodic lattices that affect the propagation of electromagnetic waves (EM) by defining allowed and forbidden electronic energy bands.[25–27] The periodic structure of a photonic crystal is built of regularly repeating internal regions of high and low dielectric constants. Several important fabrication methods were demonstrated in the literature and several commercialization efforts are already utilizing these methods, such as optical fibers,[28] artificial opals,[29] optical filters,[30] sensors,[31] and specialty pigments.[32]

Photolithography is a high quality, high cost method to fabricate two- and three-dimensional photonic crystal materials for photonic applications. This top–down fabrication concept produces complex photonic crystal materials but the high cost is likely to prevent commercialization of many photonic crystal devices.

Another important method for photonic crystal fabrication employs colloidal particle self-assembly. A colloidal system consists of two separate phases: a dispersed phase and a continuous phase (dispersion medium). The dispersed phase particles are small solid nanoparticles with a typical size of 1–1000 nanometers. Colloidal crystals are three-dimensional periodic lattices assembled from monodispersed spherical colloids. The *opals* are a natural example of colloidal photonic crystals that diffract light in the visible and near-infrared (IR) spectral regions due to periodic modulation of the refractive index between the ordered monodispersed silica spheres and the surrounding matrix.

A colloidal photonic crystal diffracts light (Fig. 4) according to Bragg's law:

$$m\lambda = 2d_{111}\, n_{eff} \sin\theta$$

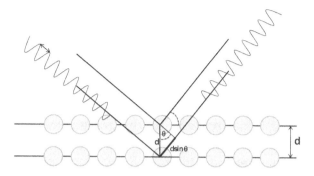

Fig. 4. Schematic illustration of Bragg diffraction from photonic crystal lattice planes.

where m is the order of diffraction, λ is the wavelength of the diffraction peak, d_{111} is the (111) lattice plane spacing, θ is the angle between the incident light and the normal to the diffraction planes and n_{eff} is the effective refractive index of the colloidal photonic crystal.

Since in photonic crystals particle size is in the submicron range, inkjet printing should be a suitable technique for obtaining patterned photonic crystals. There are indeed several reports of inkjet printing of photonic colloidal crystals.

An inkjet printing of colloidal crystals was proposed by Frese et al.,[33] describing inkjet printing processes of monodispersed particles which are able to form two- or three-dimensional photonic crystals on the substrate surface by arranging in a closely packed lattice structure on the surface. The particle size was selected so that it will diffract light in the visible spectral region, i.e., particle size of 200–500 nanometers. In this work drop-on-demand inkjet printing techniques are utilized.

Monro et al. disclosed[34] another approach that may be utilized for inkjet printing of photonic crystals. According to their approach, resinous binder and a color-effect colorant in particulate form are formulated into a coating composition. The colorant is comprised of an ordered periodic array of monodispersed particles held in a polymer, where there is a refractive index mismatch between the polymer and the particles. This ink composition will reflect light according to Bragg's law once deposited on a substrate.

Lewis *et al.* have demonstrated[35,36] a direct printing concept of colloidal gels for fabricating colloidal crystals in a woodpile structure. This method, which was demonstrated for extrusion of a continuous filament that is deposited in a layer-by-layer build sequence, can be employed for inkjet printing by designing inks with a well-controlled viscoelastic response, so that they flow through the inkjet print head and then gel and set immediately to facilitate the shape of the deposited pattern. These inks[37] are comprised of a high colloid volume fraction to minimize shrinkage during the drying process due to strong capillary forces. A model system with monosized silica spheres was coated with polyethyleneimine (PEI, MW = 2000) and suspended in water at a high concentration (46 wt%) at the pH of zero charge. Near this pH value the suspension exhibits a non-Newtonian liquid-to-gel phase transition behavior. A dramatic rise in the elastic properties of the ink is evident; the shear stress and elastic modulus increase by 4–6 orders of magnitude depending on the ink composition (see Fig. 5). The weak gel of bare silica particles had insufficient strength to support its own weight during deposition; whereas the strong gel of the PEI-coated silica could be successfully patterned to 3D periodic structures. This ink design may be extended to other types of monodispersed colloidal particles such as highly charged polymeric particles widely used for fabrication of colloidal photonic crystals.

Another approach for inkjet printing of Bragg reflectors[38] was presented by Wang *et al.* Bragg mirror is a structure consisting of an alternating sequence of layers of two optical materials. Each interface between these material layers contributes a Fresnel reflection. The layers are formed so that the phase of the reflected light waves add up constructively in a certain direction and thus the light in a specific wavelength matching the periodicity of the dielectric mirror is almost completely reflected. This approach utilizes such reflectors with a diameter allowing them to be jetted through an inkjet printer nozzle so that an array of these Bragg mirrors is printed on a reflecting screen.

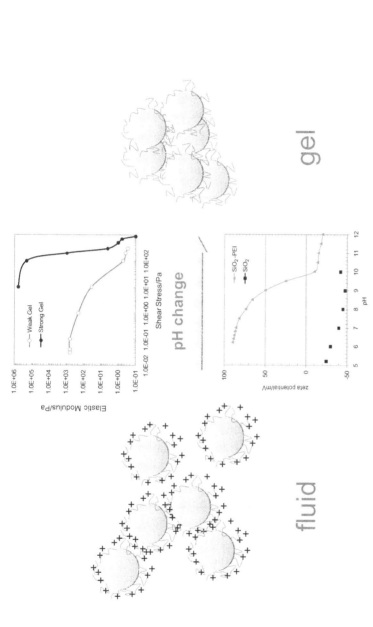

Fig. 5. Schematic illustration of the fluid-to-gel transition observed for colloidal silica inks. The bottom graph is a plot of zeta potential as a function of pH for PEI-coated silica and bare silica microspheres suspended in water. The upper graph is a log–log plot of shear elastic modulus as a function of shear stress for concentrated silica gels of varying strength: (○) denotes weak gel pH = 9.5 and (●) denotes strong gel pH = 9.75 (Ref. 36).

Inkjet Printing of Microlenses

A lens is an optical element that can focus or defocus light.[4] The most common types of lenses are circular lenses such as circular converging lenses that focus light to a point. Such lenses have many applications, such as in imaging and photography. The common circular lens has a shape that is symmetrical around the optical axis. Another important type of lens is a cylindrical lens that focuses light along a line. The typical cylindrical lens is shaped symmetrically around a principal axis, which is orthogonal to the optical axis.

Microlenses are tiny lenses with diameters as small as 10 micrometers (μm) and up to several hundreds of micrometers.[39] The small design of a microlens usually provides a way to gain high optical quality. A typical microlens may be shaped from a single element with a flat surface and one spherical convex surface to refract the light. Due to their small sizes, microlenses are usually fabricated on a substrate that supports them. A different type of microlens has two flat and parallel surfaces and the focusing action is obtained by a variation of refractive index across the lens. These are known as gradient-index (GRIN) lenses. Some microlenses achieve their focusing action by both a variation in refractive index and by the surface shape. Another class of microlens (micro-Fresnel lenses) focuses light by refraction in a set of concentric curved surfaces. Such lenses are made very thin and lightweight. Microlens arrays contain multiple lenses formed in a one-dimensional or two-dimensional array on a supporting substrate.

Historically,[40] microlenses were fabricated using a microscope. Small filaments were used to melt glass and allow the surface tension in the molten glass to form the smooth spherical surfaces required for lenses. Advances in technology have enabled microlenses to be designed and fabricated to close tolerances by a variety of direct deposition methods, such as ion exchange, chemical vapor deposition (CVD), electromigration, photolithography, diffusion polymerization and inkjet printing.

Biehl *et al.* demonstrated[41] a method to fabricate microlenses from hybrid organic-inorganic materials on glass using drop-on-demand inkjet printing with 50 μm nozzles driven by piezoelectric actuation. These lenses were prepared by hydrolysis of methacry-loxypropyltrimethoxysilane mixed with an ethanolic solution of tetraethyleneglycoldimethacylate and Irgacure 184, as a photoini-tiator. After printing, the ink drops were polymerized by UV light irradiation. This printing technique produced plano-convex spherical microlenses with diameters varying from 50 to 300 micro-meters, focal lengths from 70 μm to 3 mm and f-numbers as low as 0.6.

In a somewhat similar fashion, Ishii *et al.*[42] have demonstrated inkjet fabrication of polymeric microlenses for optical chip pack-aging. UV curable epoxy resin is deposited onto optical devices by inkjet printing. When the droplets hit the surface, they form into partial spheres due to their surface tension, and are UV-cured to form the microlens with diameters from 20 to 40 μm with f-numbers of 1.0 to 11.0. Their uniformity in a microlens array was measured to be within ±1% in diameter and ±3 μm in pitch (total count of 36 lenses). They have also demonstrated hybrid integration of inkjetted microlenses with a wire-bonded vertical-cavity-surface-emitting laser (VCSEL) with coupling efficiencies of 4 dB higher than without the microlens.

A work by Cox and Guan[43] discloses a process that utilizes inkjet printing of optical polymeric fluid to produce complex microlenses on a substrate having an axial gradient index of refraction. Two optical polymeric fluids are used, one having an index of refraction higher than the other. A base portion of the microlens is printed using the lower index of refraction material and a cap portion of the microlens is printed over the base portion to produce the microlens. Interdiffusion of the base portion and top or cap portion creates a generally uniform gradient diffusion zone in the vertical direction, where the lower boundary of the zone has the index of the base portion and the upper boundary of the zone has the index of the top portion. After a sufficient gradient diffusion zone is produced, the formed microlens is solidified by curing to stop any further dif-fusion. The microlenses may be formed as individual lenses on an optical substrate or as an array of microlenses.

Another report by Cox and Guan[44] discloses collimating micro-lenses that are printed directly on the end of optical fibers using inkjet technology. The open upper end of the optical fiber is fixed in a flat fixture and filled with optical fluid from which the microlens is formed thereon to collimate light exiting the fiber through the microlens. The fixture openings serve as a mold for the lens formation with the end of the fiber located at the focal distance of the lens formed on the fixture. A non-wetting coating can serve to control spreading of the fluid optical material and allow lens radius control. UV curable optically transparent epoxies are the preferred optical fluid for forming the microlenses.

Risen and Wang[45] developed a method and compositions for producing microlenses and optical filters. According to their method, carboxylated silicone or polysilicone precursor composition is applied to the surface of a substrate to form a precursor droplet, which is thermally oxidized to form a microlens. The substrates utilized were silica, silicates, borosilicate glasses, and silicones. The precursors, which are present in concentrated solutions, are viscous fluids which are used to form microdroplet precursors. A solvent such as ethanol or acetone is added to the precursors to modify and control their flow and surface tension properties, to facilitate the formation of spherical shape of the precursor on substrates. The precursor droplet volume is 4–600 picoliters and forms a droplet of 20 to 1000 micrometers in diameter.

REFERENCES

1. Moffatt JR. (1992) Bleed alleviation in ink-jet inks, US5116409.
2. Hair ML, Lok KP, Winnik FM. (1987) Ink jet compositions with insoluble dye complexes, US4705567.
3. Moffat JR, Bedford ET, Lauw HP. (1999) Ink-jet inks for improved print quality, US5935309.
4. Parker SP. (1998) *McGraw-Hill Concise Encyclopedia of Science & Technology*, 4th edition. McGraw-Hill: New York.
5. Breton MP, Noolandi J, Isabella M, Birkel S, Hamer GK. (1997) Ink compositions containing liposomes, EP0778322.
6. Miller R, You YS. (1991) Microemulsion ink jet ink composition, US5047084.

7. Breton MP, Wong RW, Schwarz WM, Gagnon Y, Friberg SE. (1997) Liquid crystalline ink compositions, US5643357.

8. Oliver J, Trevor I, Jennings CA, Johnson EG, Breton MP. (1996) Photochromic microemulsion ink compositions, US5551973.

9. Oliver J, Breton MP, Friberg SE, Wong RW, Schwarz WM. (1996) Liquid crystalline microemulsion ink composition, US5492559.

10. Wickramanayake P, Parazak DP. (1998) Bleed-alleviated aqueous pigment dispersion-based inkjet ink compositions, US5713989.

11. Wickramanayake P, Parazak DP. (1996) Bleed-alleviated, waterfast, pigment-based ink-jet ink compositions, US5531816.

12. Wickramanayake P, Parazak DP. (1999) Reliability enhancement of microemulsion-based ink-jet inks, EP892025.

13. Wickramanayake P. (1994) Black-to-color bleed control in thermal ink-jet printing, US5342440.

14. Tsang JW, Moffatt JR. (1998) Black-to-color bleed alleviation using non-specific ionic, pH and colloidal effects, US5853465.

15. Kabalnov AS. (2002) Microemulsion techniques for ink jet inks, US6432183.

16. Tsang JW, Moffatt JR. (1998) Preparation of microemulsion and Micellar color inks from modified water-soluble color chromaphores for thermal ink-jet printing, US5749952.

17. Wickramanayake P, Moffatt JR. (1993) Solubilization of water-insoluble dyes via microemulsions for bleedless, non-threading, high print quality inks for thermal inkjet printers, US5226957.

18. Magdassi S, Ben-Moshe M. (2003) Patterning of organic nanoparticles by Inkjet printing of microemulsions. *Langmuir* **19**: 939–942.

19. Magdassi S, Ben-Moshe M. (2003) Inkjet ink compositions based on oil-in-water microemulsion forming nanoparticles upon application on a surface, US7115161.

20. Wong RW, Breton MP, Croucher MD, Duff JM, Petroff TE, Riske W, Henseleit K. (1994) Ink-jet printing process, US5345254.

21. Caputo P. (1998) Stable oil-in-water ink emulsion based upon water-reducible nigrosine dyes for inkjet printers and felt-tip and roller-ball pens, US5746815.

22. Spinell HJ. (1998) Aqueous inkjet ink composition, US5772741.

23. Kabalnov AS. (2002) Miniemulsion techniques for ink-jet inks, US6342094.

24. Gratson G, Xu M, Lewis JA. (2004) Direct writing of three dimensional webs. *Nature* **428**: 386–389.

25. Yablonovitch E. (1987) Inhibited spontaneous emission in solid state physics and electronics. *Phys Rev Lett* **58**: 2059–2062.

26. John S. (1987) Strong localization of photons in certain disordered dielectric super lattices. *Phys Rev Lett* **58**: 2486–2488.

27. Asher S. (1986) Crystalline colloidal narrow band radiation filter, US4627689.

28. Jianzhao L, Herman PR, Valdivia CE, Kitaev V, Ozin GA. (2005) Colloidal photonic crystal cladded optical fibers: Towards a new type of photonic band gap fiber. *Opt Express* **13**: 6454–6459.

29. Nakano Y, Kamiyama K, Kobayashi T. (1987) Process for the production of jewelling and ornamental material, US4703020.

30. Asher SA, Jagannathan S. (1994) Method of making solid crystalline narrow band radiation filter, US5281370.

31. Asher SA, Holtz JH. (1998) Polymerized crystalline colloidal array *sensor* methods, US5854078.

32. Schmid R, Mronga N. (1997) Multiply coated metallic luster pigments, US5607504.

33. Frese P, Bauer RD, Egen M, Taennert K, Wulf M, Zentel R. (2005) Ink-jet ink composition, US7122078 B2.

34. Monro C, Merrit M, Lamers P. (2003) Color effect compositions, WO03058299.

35. Smay JE, Cesarano J, Lewis JA. (2002) Colloidal Inks for Directed Assembly of 3-D Periodic Structures. *Langmuir* **18**: 5429–5437.

36. Gratson GM, Garcia-Santamaria F, Lousse V, Xu M, Fan S, Lewis JA, Braun PV. (2006) Direct-write assembly of three-dimensional photonic crystals: Conversion of polymer scaffolds to silicon hollow-woodpile structures. *Adv Mater* **18**: 461–465.

37. Lewis JA. (2006) Direct ink writing of 3D functional materials. *Adv Funct Mater* **16**: 2193–2204.

38. Wang SY. (2005) Ink with Bragg reflectors, US20050094265.

39. Borrelli NF. (1999) *Microoptics Technology: Fabrication and Applications of Lens Arrays and Devices*. Marcel Dekker: New York.

40. Hooke R. (1665) *Preface to Micrographia*. The Royal Society of London.

41. Biehl S, Danzebrink R, Oliveira P, Aegerter MA. (1998) Refractive microlens fabrication by inkjet process. *J Sol-Gel Sci Tech* **13**: 177–182.

42. Ishii Y, Koike S, Arai Y, Ando Y. (2000) Ink-jet fabrication of polymer microlens, for Optical-I/O Chip Packaging. *Jpn J Appl Phys* **39**: 1490–1493.

43. Cox WR, Guan C. (2001) Inkjet printing of gradient-index microlenses, US patent application 2001048968.

44. Cox WR, Guan C. (2001) Inkjet printing of collimating microlenses onto optical fibers, US2001033712.

45. Risen WM Jr, Wang YZ. (2001) Method and compositions for producing microlenses and optical filters, US6294217.

PART III

SPECIALTY INKJET MATERIALS

CHAPTER 12

Electrically Conductive Inks for Inkjet Printing

Moira M. Nir, Dov Zamir, Ilana Haymov,
Limor Ben-Asher, Orit Cohen, Bill Faulkner*,
and Fernando de la Vega
Cima Nanotech Israel, Ltd., Caesarea, Israel
**Cima Nanotech, Minneapolis, MN, US*

INTRODUCTION

Digital printing of conductive materials has sparked significant academic and commercial interest. Much of this interest has arisen from the burgeoning fields of organic electronics and printed electronics. Research and development in printed electronics has yielded the ability to deposit electronic features on a wide range of substrates by means of standard printing technologies. Recent developments hold forth the promise of cost effective manufacture of electronic devices and displays in which all features are fully printed. With the availability of printable conductors, these goals are on their way to realization.

Conductive inkjet (IJ) inks can be employed to fabricate conductive features on a variety of substrates for a wide range of electronics applications. These features have sizes on the order of millimeters, microns, and even submicron dimensions. Alternative methods of fabricating conductive components that are currently in use waste expensive metal raw materials and generate substantial amounts of chemical pollutants. In addition, many conventional metal deposition methods do not lend themselves to automation because of their lengthy processing steps. The global market demand for high quality, low cost electronic components requires innovative fabrication techniques that are both faster and cheaper than those of traditional production methods. Inkjet printing is a digital method that offers this flexibility and cost advantage, especially for small-scale custom printing processes and automated production.

All of the advantages of general IJ processes, such as those in graphics applications, hold true as well for conductive IJ inks. Thus inkjet printing is an attractive method for direct patterning of conductive lines. At present, the emerging field of printed electronics and the growing demand for printing on flexible, temperature sensitive substrates are aggressively driving the development of conductive ink systems.

APPLICATIONS AND MARKET FOR CONDUCTIVE INKS

The market for printed electronics, including organic, inorganic, and composite materials, was estimated at $1.2 billion in 2007, with expected growth to a level of $48.2 billion level by 2017.[1]

Conductive inks, though not specifically inkjet inks, are currently used in products that make up about 83% of the (2007) printed electronics market.[1] Thick-film conductive inks and pastes are found extensively in printed circuit boards (PCBs) and medical biosensors (e.g., disposable glucose sensors for diabetic patients). The developing photovoltaic (PV) industry is the largest growing market in printed electronics, and is expected to reach over 400 million dollars by 2010 and over 6 billion dollars by 2015.[2] Most commercial PV cell configurations require the printing of a conductive pattern for the electrodes that are exposed to the light source. In the future, semiconductor materials deposited by inkjet may also be incorporated in flexible PV cells. The second largest growing market in printed electronics is for organic light emitting diodes (OLEDs). Inkjet printing of conductive materials is a viable alternative to the current fabrication processes for these products.

Another major technology area that can utilize conductive IJ ink is the display market. Inkjet can be applied for both flexible and rigid displays such as electroluminescent and electrophoretic displays (including e-paper), liquid crystal displays (LCD), plasma display panels (PDP) and touch screens; some functionalities have already been printed by IJ technology in certain display applications, for example RGB color filters. Conductive IJ is also appropriate for use in thin film transistors (TFT), disposable batteries, radio-frequency identification (RFID) tags, and a range of chemical and electronic sensors.

Most of the conductive IJ inks that are commercially available today, and a majority of the development work being performed as of 2008, utilize silver-based inks for reasons discussed later in this article. Major applications in printed electronics which are currently the focus of significant development work with conductive inkjet inks are summarized in Table 1.[3–5]

COMPARISON WITH ALTERNATIVE PRINTING TECHNOLOGIES

Photolithography has been widely adopted in the printing circuit board industry for manufacturing of circuits.[6] However, this method involves multiple steps including etching, masking, electroplating, etc., that it is time consuming, labor intensive, and expensive. Since the solvent used in the etching process is corrosive, the choice of substrate is limited. Moreover, the photolithography process generates high volumes of hazardous waste which are environmentally destructive and for which treatment is becoming prohibitively expensive. Alternative metal deposition techniques for the production of electronics parts involve vacuum deposition.

Table 1. Opportunities for conductive inkjet inks in printed electronics applications.[3–5]

Application	Incumbent Technology	Comments
Displays and backplanes	Mostly indium tin oxide (ITO).	Silver (Ag) is currently not used widely in these applications, with the exception of printed gridlines for PDPs. Large potential for Ag inks in emerging display technologies like OLEDs, electrophoretics, and cholesteric displays.
Printed PV	Thick-film Ag inks.	Currently being used to screen print-grids on c-Si solar cells and on some CIGS cells. Other PV applications under development.
RFID tags	Organic conductors, Ag thick film inks.	Thick film Ag is used for printing antennas, and organics are being developed for printed TFTs for the active portion of the tag.
Printed sensors	Organic conductors, carbon-based inks.	Under development.
Printed lighting		Under development; experimentation with Ag and organic conductors.
Printed memory		Under development; experimentation with Ag and organic conductors.
Printed circuit boards	Connectors, via filling and embedded functions.	Under development.
Smart cards/novelty consumer products/ "consumables" disposable electronics	Organic conductors.	Under development.

These methods typically require expensive initial tooling costs for parts such as masks and vacuum equipment.

In comparison, processes for printing conductive materials, specifically screen-printing and digital printing methods such as inkjet, significantly reduce costs by decreasing set-up time, by being less labor intensive, and by removing the need for high initial tooling costs. In addition, digital inkjet printing is an additive process in which materials are deposited only when and where they are needed, thereby greatly reducing system waste streams. These advantages enable frequent and less costly updates to the processes and products, potentially decreasing time-to-market for improved electronic goods.

A comparison of photolithography, vacuum deposition, and printed electronics techniques for depositing electronics features is given in Table 2 and shows the overall advantage of the printed approach, especially for up-and-coming flexible electronics applications. Although photolithography enables very high resolution which other printing methods have not yet achieved, resolution for IJ deposition is improving and has reached submicron capability when applied in combination with other methods such as substrate surface treatments.

Within the broad category of "printed electronics", there are various technologies for manufacturing conductive components, of which inkjet is one of them. These printing techniques are compared to each other in Table 3.[4,5] Inkjet is the most commercially suitable of these methods for applications where rapid customization is required and where the substrate is delicate or otherwise sensitive to direct contact.

In summary, inkjet offers non-contact digital patterning of conductive features in an additive process with the possibility of utilizing flexible and temperature sensitive substrates. The associated process involves low cost set-up and maintenance, provides relatively high resolution, and allows for quick turnover and customization. Overall, IJ printing will ultimately enable production of a broad range of electronic components that are made completely by printing processes.

Table 2. Comparison of printed metal deposition methods to photolithography and vacuum deposition.[5]

	Photolithography	Vacuum Deposition	Printed Electronics
Type of Process	Subtractive	Additive	Additive
Cost	High	Moderate	Low
Speed	High	Low	Varies according to technique
Additional Considerations	High temperature, harsh chemicals	High temperature	Low temperature
Environmental	High volume hazardous waste	Moderate volume hazardous waste	Low volume hazardous waste
Resolution	Very high	Low	Low but improving
Suitable for low cost flexible electronics	No	No	Yes
Current state of evolution	Established and possibly reaching limits of capabilities.	Widely used in coatings and electronics industry. New variants are appearing.	Emerging technology, although thick-film screen printing is well established.
Products made with technology	VLSI processor components, memory	OLEDs, coatings	EMI shielding, RFID antennas, capacitors, flexible backplanes, PV, toys and games. Other applications are emerging.

Table 3. Comparison of printed electronics technologies for conductive features.[4,5]

Printing Technology	Advantages	Disadvantages	Comments
Screen Printing	• Established technology and major ink for printed electronics today. • Inexpensive capital equipment. • Low price inks.	• Can be expensive due to amount of material required relative to yields.	• Mainly in RFID antennas and PV front electrodes. • Feature resolution ~75 microns.
Inkjet	• Can print thin films and fine features. • Amenable to prototyping. • Digital customization. • Specific equipment designs for electronics. • Non-contact method. • Able to print on wide variety of substrates. • High speed. • High quality/feature definition.	• Low throughput. • Complex systems.	• Major equipment suppliers include Fujifilm Dimatix, HP, Litrex, Xennia/Carlco, Xaar, Ricoh.
Gravure		• High set-up costs.	• Being developed for roll-to-roll printing applications.

(Continued)

Table 3. (*Continued*)

Printing Technology	Advantages	Disadvantages	Comments
Flexography	• Lower set-up costs than gravure. • Relatively good resolution.	• Prints a "donut", rather than a dot. • Susceptible to "halos". • Ink release can be inconsistent.	• Expected for RFID tags.
Offset Lithography	• High throughput. • High resolution.	• Limited in thickness (<1 micron). • Can have shearing problems. • Additives used in graphics printing can have negative effects in electronics printing.	• Not expected to be a major printing technology for electronics manufacture.
Pad Printing	• Able to print irregular configurations easily. • More adaptable to R&D environment.	• Low throughput. • Not a manufacturing-scale process.	
Spin Coating	• Good in experimental setting for testing basic properties. • Able to form very uniform film thicknesses. • Non-contact method.	• Poor resolution. • Not a commercial scale process for printed electronics.	

PRODUCERS AND PRICING

Numerous companies are working with conductive inks and related printed electronics technologies, however most of them are currently focused on screen printing inks. Conductive IJ printing is in its early stages and most of the activities in this field are still considered to be in the realm of development. Companies significantly involved in development of inkjet inks for printed electronics include Cima NanoTech (US-Israel), Cabot (US), Advanced NanoProducts (ANP, Korea), Harima (Japan), Ulvac (Japan), Novacentrix (US), Five Star Technologies (US), Kovio (US), Degussa (Germany), Conductive Inkjet Technology (UK), Xennia (UK), and Daejoo Electronics Materials (Korea). Only a few of these entities have commercially available conductive inkjet materials, among them Cima Nanotech, Cabot, Harima and ANP.[7]

Most screen printing inks and pastes that are commercially available are based on silver or silver alloy particles for reasons discussed later in this chapter. These inks and pastes have reached a mature stage for which the price is within the range of $700 to $2000 per kg silver present in the material, depending on the additional functionalities of the ink. These values can fluctuate to take into account the market price of bulk silver (currently around $550 per kg). The screen printing silver products are not suitable for inkjet applications because the particle size of the metal is well above submicron level and will cause clogging of the inkjet print head. Conductive inks for inkjet are typically based on submicron "nano" silver particles. Nanosilver particles carry a premium price ranging from $20 000 to $40 000 per kg silver that includes development of the nanotechnologies and unique production costs. These initial sampling prices are way above the targeted market price, broadly considered to be 2–3 times the price of screen printing inks (per kg silver). They are currently too expensive to promote wide development of commercial applications, and as a result, many universities and research institutes, and a large sector of the industry, are reluctant to use these "nano-inks". Thus the development of conductive IJ printed applications is evolving somewhat slowly and the industry is still in need of lower cost conductive inks for inkjet printing.

THE CHEMISTRY OF CONDUCTIVE INKS

The basic formulation requirements and properties of conductive IJ inks are similar to those of standard IJ inks such as for graphics applications, with the major exception that formulations for conductive inks contain components that allow for the most important characteristic of the printed pattern, that is, its ability to conduct electric current.

Both conductive and non-conductive IJ inks are comprised of a liquid vehicle (which determines many of the basic properties of the ink) and a dispersed or dissolved component (which gives the ink its desired functionality). In graphic inks, this component is the soluble dye or the solid pigment, and in conductive inks, it is the conductive component. For all of these cases, the ink should demonstrate compatibility with the substrate and good printability and resolution with minimum printer maintenance. The printed image must adhere well to the substrate and exhibit good durability in the anticipated light, humidity, and working conditions for which the product is expected to perform.

For the case of conductive inks, the ink must also conduct electricity after printing. This requirement typically dictates an additional sintering and/or curing step after the usual IJ printing process. Most conductive ink formulations aim to achieve as low a resistance as possible and the narrowest printed lines possible. To achieve these goals, ink properties, substrate-ink interaction, and printer capabilities must be synchronized.

Composition of Conductive Inkjet Inks

Conductive inks may be based on aqueous (water-based) vehicles or on organic solvent systems. Typical solvents for non-aqueous conductive inks are similar to the solvent systems found in graphic IJ printing. Much use is made of oxygenated solvents such as methoxypropyl acetate, ethylene glycol n-butyl ether acetate, N-methyl pyrrolidone, cyclohexanone, butyl carbitol acetate, isophorone, glycerol, ethylene glycol, propylene glycol, propylene glycol ethers, and others. Some use of propylene carbonate, toluene, methyl ethyl ketone, as well as other more exotic solvents is also known. Water soluble co-solvents may be added to aqueous systems in order to tailor pH, evaporation rate, or other properties. UV curable formulations will furthermore include initiator and monomer and/or oligomer components.

An important property in the IJ formulation is the drying rate of the ink. On one hand, faster drying after application is advantageous as it enables faster printing. On the other hand, a high rate of drying in the print head may lead to clogging of the nozzles. The drying characteristics also affect the interaction between ink and substrate. Proper choice of solvent and solvent mixtures thus helps to control and optimize drying characteristics and enables high resolution printing. In aqueous systems, the addition of humectants can serve the same function.

As for the conductive component, a variety of conductive materials have been studied in inkjet formulations, for example conductive polymers,[8–11] organometallic compounds,[12,13] metal precursors,[14] metal and metal oxide nanoparticles,[15] carbon nanotubes,[16] and even molten metals.[17–19] The use of organometal compounds or metal precursors involves subsequent reduction to metallic species but these formulations have the advantage of being in the form of a solution, rather than dispersion. In the case of molten metals, operating temperature is simply too high for commercial and flexible substrate applications. Conductive polymers have a certain advantage for flexible displays, but their resistivity is relatively high.[20] Metal and metal oxide powders are the most conductive species readily available in powder form, and commercially available conductive inkjet products often involve suspensions and dispersions formulated with metal powders, especially metal nanopowders.[6,21–24] The reasons will be discussed below.

Since the requirements for jettable inks include high reliability, i.e., minimal nozzle clogging in the print head during and between jetting actions, smaller pigment particles in an ink dispersion have an advantage over larger particles. Thus formulations that employ nano-sized particles offer a clear advantage in this regard. Wide use of metals or metal oxide nanoparticles in inkjet inks is also due to the fact that they can be produced in commercial quantities, they can be incorporated at high concentrations in dispersions, and the electrical properties that can be achieved from these formulations are quite good. Some nanoparticle producers, such as Cima Nanotech, adapt or treat the surface of the particles as part of the manufacturing process[25] in order to make them fully compatible with components and chemistry of standard inkjet inks. Physical stabilization of nanoparticle dispersions is achieved via addition of suitable surfactants to the formulation, as with non-conductive IJ formulations. However, the high density of metals, coupled with the high particle loadings that are possible with nanoparticle dispersions, may stipulate a greater overall surfactant quantity than that found in standard inkjet formulations.

In particular, silver nanoparticles and occasionally gold[26] nanoparticles are employed in inks due to their low electrical resistivity, low tendency toward oxidation, and generally high chemical stability. Other metal nanoparticles, such as copper and nickel particles, tend to oxidize and yield formulations that are less stable than silver and gold at ambient conditions. Carbon nanoparticles, which incorporate relatively inexpensive raw materials, are difficult to prepare in an industrial process and have higher resistivity than metal particles. Use of non-metal nanoparticles, such as silicon, for non-conductive electronic features, is also described in the literature on IJ inks.[27]

Silver nanoparticles and nanopowders are manufactured by a wide variety of techniques. Methods for producing silver nanoparticles can be broadly classified into the following groups:

- Mechanical methods, for example by dry milling;
- Wet chemical techniques, especially reduction and precipitation of soluble metal salts, or leaching of an alloy/oxide matrix;

- Gas-phase and aerosol methods, such as by spray pyrolysis, laser pyrolysis, plasma, or vapor phase evaporation followed by condensation.

Viscosity, surface tension, and wettability are also critical characteristics of the conductive ink formulation, just as they are for non-conductive inks. They influence printing quality by their effect on drop size and shape, drop placement accuracy, satellite formation, and wetting of the substrate. Most industrial IJ print heads utilize piezo-based technology and require ink viscosities between 8 to 25 cPs (centipoise). The surface tension of the ink is dictated by the desired interaction with the substrate and by compatibility and performance criteria of the print head as set by the print head manufacturer. Values in the range of 25–35 dyne/cm are typical. As foaming is detrimental to jetting performance, suitable anti-foaming agents may be employed to limit this phenomenon.

Inkjet formulations may also include an adhesive binder. The binder is usually a resin or resin system that is soluble or dispersed in the ink vehicle. Upon drying, the resin is heat- or UV-cured. The binder provides adhesion of the printed pattern to the substrate, imparts solvent resistance, and may protect against abrasion. Careful adjustment of the type and amount of binder is required in order to achieve proper adhesion and substrate matching. As the binder systems are mostly non-conductive, an optimum ratio between the metal and the binder system must be found. Generally, there is no one generic binder suitable for all substrates or all conditions.

Obviously, one of the major concerns in the preparation of the conductive ink is the need for low electrical resistivity. Commercial conductive IJ inks based on silver particles have metal loadings ranging from approximately 20% to 80% by weight. Resistance of a pattern printed from a single pass of a print head is thus expected to be lower for formulations with higher solids content. However, another factor that influences resistivity is the non-conductive organic load in the ink and in the sintered pattern.[12] Therefore, when choosing the additives for the formulation, care must be taken to minimize the amount of organic materials in relation to the conductive additive.

Ink Interaction with the Substrate

In general, the substrate should be inert to the solvents in the ink formulation. However, there are occasions for which the solvent is specifically chosen so as to selectively etch the substrate and thereby aid adhesion.

The interaction between the ink and the substrate, both during and after deposition, strongly influences the quality of printing. Dynamic ink-substrate interactions such as ink drop impact, wetting, spreading, penetration, and solvent evaporation rate, as well as drying and coalescence of the ink particles on the substrate, are all important factors.

The substrate surface energy must significantly exceed the surface tension of the jetted fluid in order to obtain suitable wetting of the substrate. The wetting properties of the ink on the substrate influence the minimum possible line width and the resolution between adjacent lines, as well as the homogeneity of the lines. The evaporation rate of the solvent must be slow enough to allow for wetting and drop spreading prior to subsequent drying, but not so slow that excessive spreading occurs or that drying time is unreasonable for the overall process. In order to achieve high quality patterns, when the individual ink drops are deposited on the surface they must spread, join together, and form a substantially uniform and level film. This process requires a low advancing contact angle between the ink and the substrate. For any given ink/substrate combination, the advancing contact angle is typically significantly greater than the receding contact angle. Surface energy and roughness of the substrate additionally influence the ink setting profile, ink adhesion, and rub-resistance of the printed pattern.

Substrates for conductive inkjet printing are diverse and depend on the application requirements: some are flexible polymers, especially polyimide (PI) and polyesters such as polyethylene terephthalate (PET) and polyethylene naphthelene (PEN); others are rigid polymers such as FR-4 for printed circuit boards, and some are inorganic materials such as glass or silicon. The surface of the substrate can be polished or rough, untreated or treated (e.g., by corona, plasma, chemical treatment). It may be precoated to give anti-glare or protective properties or other functionalities.

Surface treatment and substrate patterning can be employed to further reduce feature resolution. For example, drop spread can be controlled and significantly reduced by prepatterning the substrate with "dewetting patterns" that confine the ink to specific regions on the surface.[11,28] Alternatively, Smith *et al.*[29] achieved printed lines 5–30 microns thick by creating channels in the substrate and then filling them with inkjetted ink.

Adhesion of the ink to the substrate is largely dependent on the substrate type and its surface treatment and it can also depend strongly on sintering or curing conditions. Maximal curing temperature is dictated by the resistance of the substrate to heat. Both the glass transition temperature (T_g) and the melting characteristics of the substrate must be considered.

With reference to the thermal characteristics of the substrate, and in light of the proposed use of the printed pattern, there may be a need to match the thermal expansion coefficient of the substrate and the dry ink. In addition, for printing of flexible devices, the flexibility of the printed pattern must be adjusted to that of the substrate.

One of the substrates that deserves special attention in our discussion is silicon (Si), and in particular Si wafers for producing solar cells. A variety of surface treatments are applied to Si in the solar cell industry: Native Si vs. chemically protected (SiN, for example), polished versus unpolished versus patterned to give surface texture, etc. In order to ensure proper Si-metal contacts, careful control of the conductive ink composition is required. In addition, very high curing temperatures are usually employed in the manufacture of solar cells, thus there is a need for a suitable binding component. Special care is taken to avoid excessive diffusion of metal into the emitter layer, nevertheless, somewhat limited diffusion is desirable so that contact resistance is adequate.[27]

SINTERING AND RESISTIVITY

Sintering

In order to form a conductive pattern from nanoparticle inks, the potentially conductive particles must be consolidated to create a continuous percolated path for the conduction of electric current.

Sintering is defined as a process in which distinct particles in a powder weld together and interdiffuse with each other at temperatures below their melting point. The concept has been employed in the fields of powder metallurgy and ceramics for hundreds of years. Sintering allows metal particles, whether nanoparticles for inkjet applications or larger particles for other printed electronics applications, to join together at a temperature below the melt phase in order to form the conductive path.

Metallic nanoparticles on the order of a few nanometers are expected to have significantly reduced melting points and sintering temperatures relative to their bulk counterparts.[30,31] For example, the melting point of gold particles having diameters less than 5 nm is predicted to be 300–500°C lower than the that of bulk gold at 1063°C.[30] The ability to form conductive patterns at low sintering temperatures on the order of 100–300°C enables the use of temperature-sensitive substrates such as low-cost plastics in entirely new applications.

After printing a conductive ink composed of metal nanoparticles in a liquid dispersion, the solvent is removed and the remaining material is sintered in a subsequent (or simultaneous) step. During sintering, a densification mechanism occurs involving changes in the size and shape of both the particles and the pores between the particles.[32] Specifically, sintering involves three stages: in the initial stage, the particles are in point contact. "Neck growth" occurs between the particles such that the contact area between them increases and the overall surface area decreases slightly. The original particles are still distinguishable and there is extensive interconnected "pore" space within the material, with no significant change in the overall material density. In the intermediate stage, necking expands, grain boundaries between the original particles migrate due to grain growth, and the original particles are no longer distinguishable, nevertheless, the pores are still more or less inter-connected throughout the material. The final stage is identified by isolation of spherical shaped pores with reduced size and number, and by maximal grain growth that leads to the absence of grain boundaries. In this stage there is significant material shrinkage and a corresponding increase in density.

The sintering step can be accomplished by exposure to heat,[6,21] laser,[22] pressure, photo irradiation, microwave radiation,[33] or plasma.[34] Parameters affecting the sintering process include temperature, time, particle size, particle shape, energy level of applied radiation, and in some cases, the thickness of the printed pattern. There is also an indication in the literature that aqueous formulations allow for lower sintering temperatures relative to formulations with organic solvent as the vehicle.[35] Decreasing particle size promotes sintering due to the increased surface-to-volume ratio and the augmented surface energy.[33] The degree of sintering increases with temperature and to a lesser extent also with time, therefore the sintering or curing temperature schedule selected for a thermal process will highly influence the resistivity of the conducting pattern. On the other hand, the thermal limitations of the substrate must also be considered, as well as the influence that over-heating may have on the adhesion of the ink to the substrate.

To give specific examples, an inkjet printed silver nanoparticle ink reported by Lee *et al.*[6] employed a water-diethylene glycol solvent system which was sintered at 260°C for 3 minutes and gave a resistivity of 16 $\mu\Omega$ cm, only ten times greater than that of bulk silver at 1.6 $\mu\Omega$ cm. Fuller *et al.*[21] studied a silver nanoparticle inkjet ink consisting of 5–7 nm particles dispersed in α-terpineol. After sintering at 300°C for 10 min, a printed silver line was found to have a resistivity of approximately 3 $\mu\Omega$ cm. Bieri *et al.*[22] utilized a toluene suspension of gold nanoparticles with average size of 2–5 nm, and following inkjet printing, an argon laser beam was focused on the printed line to evaporate the solvent and then sinter the particles to give conducting lines with resistivity of 14 $\mu\Omega$ cm, approximately six times greater than that of bulk gold (2.2 $\mu\Omega$ cm). Perelaer *et al.*[33] applied microwave radiation to a suspension of silver nanoparticles with average size of 5–10 nm in tetradecane inkjet printed on polyimide. The resistivity obtained for the sample was 30 $\mu\Omega$ cm.

Determining Resistivity of Conductive Inks

We mentioned above that the resistivity of bulk silver is 1.6 $\mu\Omega$ cm (at 20°C), and for gold it is 2.2 $\mu\Omega$ cm. But what is "resistivity"?

Volume Resistivity of Solid Conducting Materials

Electrical resistivity, ρ, is an inherent property of a bulk material, and is the reciprocal of conductivity, σ. Resistance is related to resistivity as follows:

$$R = \rho \cdot L_c/A_c \qquad (1)$$

where R = measured resistance, ρ = volume resistivity, A_c = conductive cross-sectional area, and L_c = conductive path length. Volume resistivity for a solid conductive material can thus be back-calculated from measured resistance data and the dimensions of the sample by rearranging Eq. 1:

$$\rho = R \cdot A_c/L_c \qquad (2)$$

Current Practice for Resistivity Determination
of Printed Inks

Reported volume resistivities for printed patterns formed from commercial silver-based inks are higher than that of bulk silver. This occurrence reflects the fact that sintered ink patterns contain non-ideal defects such as incomplete particle-to-particle contact, incomplete sintering between contacting particles, residual porosity, and the presence of non-conductive additives. The morphology and extent of void formation in two representative sintered silver nanoparticle inkjet inks are illustrated in Fig. 1.

Current commercial practice for evaluating volume resistivity of printed conductive material relies on the overall dimensions of the test specimen, such as in the ASTM F1896 standard "method for

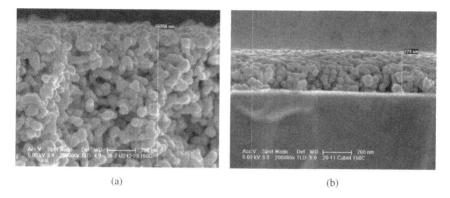

(a)   (b)

Fig. 1. Morphology of nanosilver inks after sintering. High resolution scanning electron micrographs of cross-sections of inkjet printed features after sintering at 150°C for 60 minutes: (a) Cima Nanotech ink, (b) Cabot ink.

determining electrical resistivity of a printed conductive material". Resistivity is typically evaluated by the following procedure:

- A conductive ink is coated or printed in a simple pattern having length L and width B and is then sintered according to manufacturer's specifications.
- Resistance R of the patterned test area is measured by a 4-point probe and total sample height, h_t, is determined by direct measurement, for example by profilometry or microscopy.
- Resistivity ρ_t is calculated as

$$\rho_t = R \cdot A_t/L = R \cdot h_t \cdot B/L \qquad (3)$$

Where A_t is the total cross sectional area of the sample. When resistance is measured for the case of B = L, then R is referred to as R_{sq}, (or $R_\square$ or R/sq or "sheet resistance") and Eq. (3) simplifies to

$$\rho_t = R_{sq} \cdot h_t \qquad (4)$$

Thus in commercial practice, volume resistivity of a printed pattern is not calculated from actual conductive area A_c and conductive path length L_c, rather, it is calculated using the readily measured dimensions of total sample height (h_t), or total cross sectional area (A_t) and sample length (L). Since the printed pattern typically contains void space and/or non-conductive material, and since the conductive path is not completely unidirectional, A_t is larger than the actual conductive area A_c, and L is shorter than the actual conductive path L_c. Therefore, the value of ρ calculated is higher than that obtained if values corresponding to the conductive area A_c and conductive path length L_c are used. The value ρ_t thus represents a proportionality constant relating the measured resistance to overall sample geometry, rather than an inherent property of the conductive portion of the material.

An open question is therefore incumbent on the industry as to which method is preferred for calculating resistivity of printed patterns: One in which the cross-sectional area A_t is based on total volume, or an alternative method in which only the solid conductive cross-sectional area A_s is considered, for example a mass-based method. The latter approach is also valid since only the solid conductive portion of the sample has the potential to conduct electricity.

Volume Resistivity by Mass-based Approach

Mass-based resistivity ρ_s for the same pattern above with resistance R is determined as:

$$\rho_s = R \cdot A_s/L \qquad (5)$$

Here A_s is the equivalent conductive cross-sectional area of the sample and is related to the mass of the conductive fraction of the pattern ($mass_c$) and density (d_s) of the solid conductive material, (e.g., 10.45 g/cm^3 for silver):

$$A_s = (mass_c/d_s)/L \qquad (6)$$

Values obtained by the two methods are related as follows:

$$\rho_s/\rho_t = A_s/A_t$$

which also corresponds to the ratio of non-conductive volume relative to conductive volume of the sintered ink. Thus ρ_s is smaller than ρ_t except for the case of a perfectly sintered mass without voids and free of non-conductive material, in which case ρ_s and ρ_t are equal.

PRINTING ISSUES

Inkjet is a non-contact additive process that delivers metered amounts of a variety of fluids at a precise location in time and space. It requires low viscosity liquid phase inks that must successfully be fired through a nozzle. Inkjet has broad appeal due to its seemingly simple nature. Although all inkjet processes share basic features, the ways in which inkjet technology can be configured are endless.[36] For the most part, drop-on-demand printers have been used in inkjet printing of electronics.[7]

Resolution of Inkjet Printing

Inkjet printing is a matrix printing method. Resolution defines the center-to-center dot spacing on substrate; this spacing, together with the degree of dot overlap desired, governs the choice of dot diameter.

State-of-the-art piezo-based drop-on-demand printers are produced by companies such as Xaar (UK), Konica Minolta (Japan), and Fujifilm Dimatix (US). These print heads have around 180 to 360 nozzles per inch and are designed to print at basic resolutions of up to 360 dpi with drop volumes down to about 40 pl (picoliter). This type of print resolution gives features equivalent to approximately 100 μm track width on PET or PI substrates. However, Xaar's new generation "gray scale" heads allow variable drop volumes down to around 3 pl to give printed features of around 50 μm or less.

In the academic setting, inkjet resolution reported for printed organic electronic materials is approximately 20–25 μm for untreated substrates and 5 μm for thermal laser printing.[11] For conductive metallic features, features of 10 μm or less have been reported for untreated surfaces.[37] These limits can be further reduced by pretreating the substrate, for example by creating hydrophobic/hydrophilic dewetting patterns.[11,28,29]

Quality in matrix printing depends strongly on the accuracy with which individual dots are placed on substrate. Vertical placement errors are primarily caused by the inkjet system and horizontal errors by print head velocity variations. Dot placement errors fall into two general classes: Those that affect all dots uniformly (either vertically or horizontally), causing size variations either in height or width of the character being printed, and those that affect the placement of individual dots relative to neighboring dots. Misplacement of dots relative to adjacent drops can cause raggedness of the edges, voids, and non-uniform pattern height. These issues are especially problematic in places where the printed pattern is not straight, for example bends and corners. Incomplete filling of conductive patterns detrimentally affects the resistance obtained, as discussed below.

Morphology of the Printed Pattern

There are two critical factors that have a significant effect on the morphology of printed pattern. One factor is that of ink properties, where those properties are viscosity, surface tension, solids loading, wettability, and dispersion stability. In general, printing with a high contact angle, high viscosity, and high tension ink produces smaller sized dot or line patterns as compared to the ink that has opposite properties.[38–40] The other factor is the selection of printing parameters, such as volume and traveling velocity of the ejected droplet as determined by the wave form of the voltage signal, and gap distance between each droplet, and printing frequency. Generally, volume and velocity of jetting droplet increase with higher applied voltage and volume is maximized at optimal duration time.[41]

Unlike the mature industry of graphic printing, conductive inkjet printing is an emerging technology that faces many challenges. Among the challenges of printing conductive materials is obtaining the desired morphology and pattern geometry. The ability to control the morphology of the printed pattern plays an important role in determining its electrical resistivity and mechanical adhesion property. In general, the printed patterns must meet strict geometric requirements — no missing spots, a smooth surface, and very well-defined and straight borders. Any deviation from the above requirements will adversely affect the electrical properties, from open-circuit conditions in the case of incomplete coverage, to changes in resistance due to the deposition of less or more metal in a certain part of the pattern. Printing a less-than-desired amount of conductive material may even cause overheating due to higher than designed resistances.

The height of a printed line is not typically uniform along the printed width, and a cross-section of the line along its width may show raised edges due to the "ring stain effect".[42] A high concentration of solute in the ink is reported to reduce this effect.[12,35,43] Special attention must also be taken when printing curves and non-linear patterns such as bends, corners, and other complex geometries, as these are harder to print with inkjet heads while at the same time achieving the above requirements. For some applications, it will be necessary to develop strategies and materials to enable multilayer printing, i.e., layers of the same or of different materials. For example, since the resistance of a given pattern depends on the resistivity of the sintered material and the amount of conductive material deposited, there are cases where it is necessary to make multiple passes to build the height of the feature and obtain the desired overall resistance. Conductive patterns with a non-uniform surface are generally unsuitable for use in multilayer interconnecting structures, since subsequent layers are prone to pin holes due to the poor coverage of the numerous ridges and valleys.[26]

PROCESS INTEGRATION FOR PRINTING SYSTEMS

The complete system for manufacturing a desired product with printed features includes the printing and maintenance equipment, software, inks, substrates, conveying equipment, user-interface software, and all other application-specific requirements.

At present, no single manufacturer provides all of the components for every application. Print head suppliers provide electronics, software and technical support based on their specific experience. Ink and substrate suppliers can provide information on expected properties, adhesion, and ink stability, and perhaps optimum conditions for jetting on selected substrates. Equipment manufacturers may or may not have experience with the specific application to provide extensive technical support.

Thus there is a need for process and equipment integration for automated inkjet printing systems of the future. This integration must include handling and conveying of materials into and out of the printing device, motion control hardware and software to position the print head and substrate relative to each other, a curing or drying system to fix the jetted fluid to the substrate and sintering equipment to obtain a conductive pattern.[36] All of the elements in the system must be synchronized to perform at the same rate and speed for the case of a continuous process. Companies that are starting to develop integrated systems for inkjet operations include ITRI (Taiwan), DGI, Printar (Israel), Litrex (USA/Japan), Epson (Japan), Inca (UK), Unijet (Korea), and Konica Minolta (Japan), among others.

Two examples that clearly illustrate the need for an integrated process approach are active RFID tag printing and solar cell printing. In order for RFID tag printing to be efficient and economical, the overall process must be able to not only print the antenna but also to handle, position, attach, and connect the active chip. In solar cell printing, it is necessary to convey the cell wafer into and out of a printing system which must also include a sintering step. These wafers are typically very thin and fragile, and they make up a large fraction of the overall cost of the solar cell. Since wafer breakage will make the process uneconomical, the handling and conveying system must be gentle and accurate. Although there are less expensive printing technologies for solar cells and other electronics components, inkjet printing has financial merit when the economics of an overall integrated system are considered.

An integrated process approach for inkjet applications is not only a substitute for existing processes but also a method that enables the creation of new products and markets. While this approach is one of the biggest opportunities for inkjet printed electronics, it also has some significant barriers, those being a slow development curve for new markets, the lack of established standards, and the need for technical cooperation between different independent companies.[44] Furthermore, there is the drawback that no single market has yet taken the initiative to lead this effort. ITRI (Taiwan) has estimated that there is a demand for over 4000 inkjet printing platforms for various different applications.[45] This estimate illustrates the current lack of market focus for process integration, the type of focus that is needed to concentrate development resources.

FUTURE TRENDS

The current high price of commercially available conductive inkjet inks is hindering development of new applications, thus there is a demand for less expensive solutions. R&D efforts will continue to be devoted to development of jettable conductive inks, and new processes and/or cheaper materials, such as copper[24,43] or organic conductors, are expected to find their way into this market.

In the near future, we expect to see adoption and adaptation of inkjet technology to rapidly growing markets such as photovoltaics, OLEDs, and thin film transistors. In general, the development of flexible printed electronics for a wide range of applications is expected to take advantage of the value proposition of inkjet printing of conductive inks.

The special properties of nanomaterials, such as those currently used as the conductive "pigment" in conductive inkjet inks, will be exploited more and more in advanced second-generation products. New inks will be developed to print additional electronic function-alities such as resistors, capacitors, and semiconductors, thereby enabling inkjet printing of complete electronic components and devices. Inkjet's capability to layer materials and build 3D structures will be further developed for microelectronics applications.[7] These efforts have already been initiated in universities, research institutes, and commercial entities — in the large established ink companies as well as smaller startups financed by joint VC and government funding.

Further down the road, we will also see more combinations of existing technologies for developing and implementing new approaches in the field of printed electronics. For example, soft lithography combined with inkjet technology has a great potential for expanding the scope of inkjet.

We will see more companies developing and utilizing integrated printing systems that will match process speeds of conventional printing systems and combine inkjet advantages such as non-contact printing, high accuracy and resolution, and cost reduction via drop-on-demand technology. These systems will also offer sophisticated driving systems and software which will be able to accurately print patterns, align patterns, and precisely overlay multiple layers and features. Via filling by inkjet is then expected to follow in order to meet a full set of processing requirements. Many companies are already pursuing these goals, for example Unijet (Korea), iTi (USA), Printar (Israel), DIP-Tech (Israel), Xennia (UK), Epson (Japan), and PixDro (Netherlands-Israel). Others may follow in the future, such as the major print head producers Fujifilm Dimatix (US), Xaar (UK), and Ricoh (Japan).

A number of academic institutes are taking the initiative and developing completely functional systems and devices using inkjet tools. Some are designing their own tools to perform this basic research, thereby allowing additional degrees of freedom to their research. The rationale behind this trend may be found in the following two points:

1. These new tools and technologies and the associated freedom of design are opening up a new world of capabilities, as well as new properties, some of which will by necessity lead to the revision of outstanding theories and will increase our understanding of processes in the fields of materials, fluidics, electronics, etc.
2. There is an increasing need on the part of universities to focus at least some of their efforts in fields for which generous funding is available.

As these processes continue to develop, we also expect to see new modes of cooperation between academia, research institutes, and industry. New and innovative intellectual property (IP) arrangements may be necessary to protect the IP in a fair manner while enabling quick technology transfer and commercial implementation of the developments. Some of the existing legal approaches to these issues are somewhat obsolete and do not match present day requirements for the rapid growth of this high capital industry.

In conclusion, the huge and rapidly growing markets for printed electronics are providing the motivation for new developments in the field of conductive inkjet materials and technologies. In particular, conductive inkjet inks will find their place in new generation photovoltaics, OLEDs, and flexible consumer products. Inkjet printing of electronics will continue to offer tremendous opportunities in academia and industry in order to solve problems, to improve existing applications, and even to create new applications enabled by novel techniques and approaches.

REFERENCES

1. Das R. (2007) IDTechEx Report: Organic & printed electronics forecasts, players & opportunities. http://www.idtechex.com/printed-electronicsworld/articles/organic_and_printed_electronics_forecasts_players_and_opportunities_00000640.asp.

2. Harrop P. (2007) The global situation with printed and thin film photovoltaic beyond silicon, IDTechEx Conference — Printed Electronics Asia 2007.

3. Creagh LT. (2004) Ink Jets as Tools for Digitally Printing Electronics, Printed Electronics 2004 Conference, New Orleans.

4. Gasman L. (2007) Silver powders and inks for printed electronics: 2007–2014, NanoMarkets, LLC.

5. Gasman L, Casatelli L. (2007) Printed electronics: A manufacturing technology analysis and capacity forecast, NanoMarkets, LLC.

6. Lee HH, Chou KS, Huang KC. (2005) Inkjet printing of nanosized silver colloids. *Nanotechnology* **16**: 2436–2441.

7. Blum JB. (2007) Ink jet printing for high-frequency electronic applications, *Printed Circuit Design & Fab*, October 2007, http://pcdandf.com/cms/content/view/3839/95/

8. Kawas T, Sirringhaus H, Friend RH, Shimoda T. (2000) IEEE International Electron Devices Meeting, December, 2000.

9. Sirringhaus H, Kawas T, Friend RH, Shimoda T, Inbasekaran M, Wu W, Woo EP. (2000) *Science* **290**: 2123.

10. Morrin A, Crowley K, Ngamna O, Wallace GG, Killard AK, Smyth MR. (2007) http://spring07.ise-online.org/files/abs061689.pdf.

11. Bao Z. (2004) Conducting polymers fine printing. *Nat Mater* **3**: 137–138.

12. Cuk T, Troian SM, Hong CM, Wagner S. (2000) Using convective flow splitting for the direct printing of fine copper lines. *Appl Phys Lett* **77**: 2063.

13. Rozenberg GG, Bresler E, Speakman SP, Jeynes C, Steinke JHG. (2002) Patterned low temperature copper-rich deposits using inkjet printing. *Appl Phys Lett* **81**(27): 5249.

14. Liu Z, Su Y, Varahramyan K. (2005) Inkjet-printed silver conductors using silver nitrate ink and their electrical contacts with conducting polymers. *Thin Solid Films* **478**: 275–279.

15. Calvert P. (2001) Inkjet printing for materials and devices. *Chem Mater* **13**: 3299–3305.

16. Krisztián K, Mustonen T, Tóth G, Jantunen H, Lajunen M, Soldano C, Talapatra S, Kar S, Vajtai R, Ajayan PM. (2006) Inkjet printing of electrically conductive patterns of carbon nanotubes. *Small* **2**: 1021–1025.

17. Liu QM, Orme J. (2001) High precision solder droplet printing technology and the state-of-the-art. *J Mater Process Technol* **115**: 271–283.

18. Orme M, Smith RF. (2000) Enhanced aluminum properties by means of precise droplet deposition. *ASME J Manuf Sci Eng* **122**: 484–493.

19. Gao F, Sonin AA. (1994) Precise deposition of molten microdrops: The physics of digital microfabrication. *Proc R Soc Lond A* **444**: 533–554.

20. Winter I, Reese C, Hormes J, Heywang G, Jonas F. (1995) The thermal ageing of poly(3,4-ethylenedioxythiophene). An investigation by X-ray absorption and X-ray photoelectron spectroscopy. *Chem Phys* **194**: 207.

21. Fuller SB, Wilhelm EJ, Jacobson JM. (2002) Inkjet printed nanoparticle microelectromechanical systems. *J Microelectromech Syst* **11**: 54–60.

22. Bieri NR, Chung J, Poulikakos D, Grigoropoulos CP. (2004) Manufacturing of nanoscale thickness gold lines by laser curing of a discretely deposited nanoparticle suspension. *Superlattices Microstruct* **35**: 437–444.

23. Redinger D, Molesa S, Yin S, Farschi R, Subramanian V. (2004) An inkjet-deposited passive component process for RFID. *IEEE Trans Electron Dev* **51**: 1978–1983.

24. Volkman SK, Pei Y, Redinger D, Yin S, Subramanian V. (2004) Ink-jetted silver/copper conductors for printed RFID applications. *Mater Res Soc Symp Proc* **814**: 17.8.1.

25. Garbar A, de la Vega F, Matzner E. (2006) Method for the production of highly pure metallic nano-powders produced thereby, US20060112785 patent publication.

26. Molesa S, Redinger DR, Huang DC, and Subramanian V. (2003) High-quality inkjet-printed multilevel interconnects and inductive components on plastic for ultra-low-cost RFID applications. *Mater Res Soc Symp Proc* **769**: H8.3.1–H8.3.6.

27. Curtis CJ, van Hest M, Miedaner A, Kaydanova T, Smith L, Ginley DS. (2006) Multi-layer inkjet printed contacts for silicon solar cells. *IEEE Proceedings, IEEE Conference on Photovoltaic Energy Conversion (WCPEC-4)* **3**: 1392–1394. http://www.nrel.gov/docs/fy06osti/39902.pdf.

28. Wang JZ, Zheng ZH, Li HW, Huck WTS, Sirringhaus H. (2004) Dewetting of conducting polymer inkjet droplets on patterned surfaces. *Nat. Mater* **3**: 171–176.

29. Smith PJ, Hendriks C, Perelaer J, Schubert US. Conductive silver tracks produced by hot-embossing and inkjet printing. http://www.parthen-impact.com/parthen-uploads/add26321.pdf.

30. Buffat P, Borel JP. (1976) Size effect on the melting temperature of gold particles. *Phys Rev A* **13**: 2287–2298.

31. Lai SL, Guo JY, Petrova V, Ramanath G, Allen LH. (1996) Size-dependent melting properties of small tin particles: Nanocalorimetric measurements. *Phys Rev Lett* **77**: 99–102.

32. Arcidiacono S, Bieri NR, Poulikakos D, Grigoropoulos CP. (2004) On the coalescence of gold nanoparticles. *Int J Multiphas Flow* **30**: 979–994.

33. Perelaer J, Gans BJ, and Schubert US. (2006) Ink-jet printing and microwave sintering of conductive silver tracks. *Adv Mater* **18**: 2101–2104.

34. Renn M. (2007) Method and apparatus for low-temperature plasma sintering, WO2007/070868 patent publication.

35. Smith PJ, Shin DY, Stringer JE, Derby B, Reis N. (2006) Direct ink-jet printing and low temperature conversion of conductive silver patterns. *J Mater Sci* **41**: 4153–4158.

36. Mills RN and Demyanovich WF. (2005) Materials and process development for digital fabrication using ink jet technology. *Digital Fabrication 2005* **1**: 8–12, Society for Imaging Science and Technology, Virginia. http://www.iticorp.com/pdf/digfab_05_v0.3.pdf.

37. Murata K, Matsumoto J, Tezuka A, Matsuba Y, Yokoyama H. (2005) Super-fine ink-jet printing: Toward the minimal manufacturing system. *Microsystems Technology*, **12**: 2–7.

38. Wu X, Phan-Thien N, Fan X, Ng TY. (2003) A molecular dynamics study of drop spreading on a solid surface. *Phys Fluids* **15**: 1357–1362.

39. Furbank RJ, Morris JF. (2004) An experimental study of particle effects on drop formation. *Phys Fluids* **16**: 1777–1790.

40. Bussmann M, Chandra S, Mostaghimi J. (2000) Modeling the splash of a droplet impacting a solid surface. *Phys Fluids* **12**: 3121–3132.

41. Reis NC, Ainsley CB, Derby B. (2005) Ink-jet delivery of particle suspensions by piezoelectric droplet ejectors. *J Appl Phys* **97**: pp. 094903-1 to 094903-6.

42. Magdassi S, Grouchko M, Toker D, Kamyshny A, Balberg I and Millo O. (2005) Ring stain effect at room temperature in silver nanoparticles yields high electrical conductivity. *Langmuir* **21**: 10264–10267.

43. Park BP, Kim D, Jeong S, Moon J, Kim JS. (2007) Direct writing of copper conductive patterns by ink-jet printing. *Thin Solid Films* **515**(19): 7706–7711 (Proceedings of Symposium I on Thin films for large area electronics, MERS 2007).

44. IT Strategies (2007) Ink jet technology as a manufacturing process: Much promise, but a long way off (web). http://www.it-strategies.com/news/59.htm.

45. Cheng K. (2007) Technology progress and opportunities of printed flexible displays & electronics in Taiwan, IDTechEx Conference — Printed Electronics Asia 2007.

CHAPTER 13

Inkjet 3D Printing

Eduardo Napadensky

Objet Geometries Ltd., Israel

INTRODUCTION

Inkjet three-dimensional (3D) Printing is a fast, flexible and cost effective technology which enables the construction of both simple and complicated 3D objects directly from Computer-Aided Design (CAD) data without the need for tooling. Inkjet 3D Printing belongs to a broader family of manufacturing technologies known as Solid Freeform Fabrication (SFF).

In order to understand what is Inkjet 3D Printing and its impact on the state-of-the-art industry, it is first necessary to understand the extent and the types of changes the manufacturing industry is presently undergoing.

In the last couple of decades we have been witness to the emergence of SFF revolutionary manufacturing technologies. Today, SFF allows fast and economic building of complicated structures, which were difficult or sometimes even impossible to manufacture in the past.

SFF facilitates rapid fabrication of prototypes without the need for tooling, providing rapid and effective feedback to the designer, thus shortening the product development cycle and significantly improving the design process. SFF is mainly employed in design related fields, where it is usually used for visualization, demonstration and mechanical prototyping. SFF is also used for rapid fabrication of non-functional parts, e.g., for assessing various aspects of a design such as aesthetics, fit, and assembly. Additionally, SFF has proven to be useful in the medical field, where the outcomes are anatomic models which are built prior to a surgical procedure, thus enabling detailed preplanning of the surgical procedure, in order to shorten the process duration and reduce uncertainties. An example of a model of a human anatomic part is shown in Fig. 1.

It is well recognized that in the future, many other areas will also benefit from SFF technologies, including the fields of architecture, dentistry, and tissue engineering.

Although different types of SFF technologies have already been adopted in vastly different areas, it is believed that the revolution is only in its infancy. To date, SFF has been adopted mainly as a rapid prototyping (RP) technology, but it is expected that the next step will be rapid manufacturing (RM) of final products.

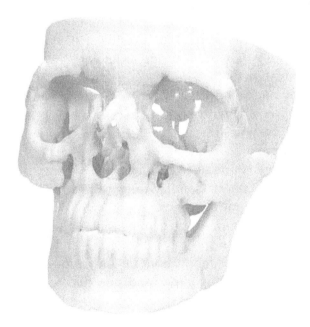

Fig. 1.   Human skull as an example of an anatomic model built by PolyJet 3D Printing.

Today there are only a few instances of SFF used as a rapid manufacturing technology; one of them being rapid manufacturing of hearing aids.

The reason why broader adoption of SFF technologies for rapid manufacturing of final products is impeded resides in deficiencies found in each of the available SFF technologies. For example, to date no single SFF technology enables fast manufacture of final products with sufficient accuracy, surface quality, and appropriate material mechanical properties, without requiring an extensive investment in labor. Although some SFF technologies enable the use of relatively high performance materials, e.g., Fused Deposition Modeling (FDM) and Selective Laser Sintering (SLS), for many applications, the resulting products lack appropriate accuracy and surface quality.

One of these new SFF technologies which aspires to reach a leading industry position is 3D Printing. Amongst the different existing SFF technologies, in recent years Inkjet 3D Printing has experienced one of the fastest market penetration rates. It is expected that the development of improved materials will enhance this rate even more.

SOLID FREEFORM FABRICATION

Solid Freeform Fabrication (SFF) technologies have in common the capability to build objects with complicated 3D geometries, directly from Computer-Aided Design (CAD) files, without the necessity of tooling. SFF technologies also have in common the employment of an *additive building process*, in which a 3D object is built up by the repeated addition of layers of material. This is in contrast to more standard technologies which are characterized by the implementation of a *subtractive fabrication process*, e.g., CNC (Computer Numerical Control), where the 3D object is manufactured by calculated removal of material from a block of raw material.

In SFF technologies, the additive building process may be divided into two steps. In the first step, a software data file which contains a virtual representation of the desired 3D object is manipulated. This manipulation enables transformation of the original file into a set of data files, each containing the virtual representation of a different single slice from the real object; the thinner the slices, the more accurate the virtual representation of the object. This step is called the "Slicing" process.

In the second step, according to the resulting slice data representation, the object is physically built up, one slice at a time, one on top of the other, in a layer by layer process, sometimes called "Recoating". These steps are schematically shown in Fig. 2.

SFF encompasses many different approaches to additive fabrication, including Stereolithography (SLA), Selective Laser Sintering (SLS), Electron Beam Melting (EBM), Laminated Object Manufacturing (LOM), Fused Deposition Modeling (FDM), and 3D Printing.

SFF technologies may be characterized by the specific process used to deposit material layers, namely the Recoating process, by the building materials used to build the 3D object itself and by the material and method used to support the object during the SFF building process:

- **Recoating process**: Recoating is the name given to the process of material layer deposition during the SFF building process, e.g., the spreading of liquid or powder as in Stereolithography and Selective Laser Sintering, respectively. An additional example is Inkjet technology, which is used for the selective deposition of layers of material in the 3D Inkjet Printing process.
- **Building material**: Different SFF technologies use different types of building materials, e.g., FDM uses thermoplastic materials,

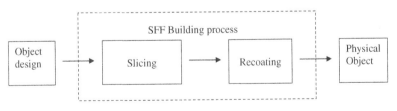

Fig. 2. Schematic representation of SFF process step sequence allowing the build-up of the physical 3D object.

Stereolithography and PolyJet 3D Printing use UV curable materials.

- **Support**: During the 3D object building process, in order to enable building of under-cut areas, different SFF technologies use different methods and materials to support the 3D object. In SLS, for example, the same material used to build the desired 3D object is used to support the object while it is being built. In this technology, at the end of the building process, the desired 3D object is immersed in non-sintered powder. As a matter of comparison, in PolyJet 3D Printing technology, two materials are used, one for building the desired object and the other one for building a support construction. The support construction is subsequently removed at the end of the building process.

At present, several SFF technologies have been developed. In the following section, those more widely accepted will be briefly presented.

INKJET 3D PRINTING

There are several methods that fall into this category.

MIT Method[1]

In this method the slices are produced by first spreading layers of powdered material onto a surface, followed by selective deposition of a liquid composition onto the powder layer surface. The liquid composition, which is deposited by Inkjet, binds the powder particles together to form a solid but fragile material.

Although this method provides a relatively fast and inexpensive building process, the resulting object generally suffers from a rough surface quality.

In addition, in existing systems, in order to achieve acceptable material mechanical properties, it is necessary to impregnate the resulting objects using a reinforcing resin. This last condition makes this process less straightforward than others.

Thermal Phase Change[2]

In this technique, a wax-like material, which is solid at room temperature, is melted and deposited by inkjet onto a substrate. The wax solidifies upon cooling, before a new layer of liquid wax is deposited. Support constructions are built in the same way as model construction layers and use the same material.

The main disadvantages of this technique are the relatively poor surface quality, difficult support removal and the poor mechanical properties of the wax-like materials used.

PolyJet[3]

In this method, the slices are formed, layer by layer, by a process of selective Inkjet deposition of UV curable liquid compositions and immediate solidification by exposure to flood UV radiation. The support construction is built in the same way as the slices of the 3D object, but using a different material. The support material is generally a UV curable gel-like material which enables easy removal and leaves a well-defined and smooth object surface.

Because PolyJet has proved to overcome many of the traditional disadvantages of existing 3D printing approaches, this method will be exposed in greater detail in the next section.

Stereolithography[4]

Although Stereolithography (SLA) is not an Inkjet 3D printing process, this method is briefly described here in order to give a better perspective on the state-of-the-art SFF technologies.

In this technique, layers of liquid UV curable material are alternately spread and selectively cured by a laser which scans the surface according to specific slice size and shape. The laser causes the liquid layer to polymerize wherever the laser beam strikes the surface, resulting in the creation of a solid plastic layer just beneath the surface.

This process of recoating and curing is repeated until the desired slice thickness is achieved. Depending on the slice thickness, each slice may have the same or greater thickness as each recoating layer.

After building one slice, a new slice is built over the previous one until the entire object is built.

In SLA, support constructions are built using the same material used to build the desired 3D object. This represents probably one of the main drawbacks of this technology. The use of a single material to build the desired 3D object and support construction results in the need to invest relatively extensive labor in order to remove the support structure, as well as for object finishing, e.g., polishing.

Selective Laser Sintering[5]

Despite the fact that Selective Laser Sintering (SLS) is not an Inkjet 3D printing process, due to its industrial importance, a brief description of this method is included here.

In this technique, slices are produced by the repeated process of spreading layers of a powdered material and selectively fusing parts of each layer by laser beam. Here, the non-fused powder serves as support, resulting at the end of the building process, an object immersed in the non-fused powder.

Probably the main disadvantages of this technique are the relatively rough surface quality obtained and the relatively high apparatus cost.

POLYJET PROCESS

PolyJet was developed by Objet Geometries Ltd. at the end of the 1990s, for the first time enabling construction of 3D objects directly from CAD data, using a very easy, straightforward, cost effective and safe process.

PolyJet systems work very similarly to standard two-dimensional (2D) printing systems. Instead of using a set of colored inks, UV curable building materials are used: One mainly for the building of the desired 3D object (the Modeling material) and one mainly for supporting of undercut areas (the Supporting material).

Immediately after building material deposition, the newly deposited layer is briefly exposed to UV radiation from a flood

UV radiation source, before a subsequent layer is deposited. At the end of the building process, the cured Supporting material is removed manually or with the aid of water.

When comparing PolyJet technology to previously existing rapid prototyping technologies such as SLA, it is immediately apparent that the approach adopted by PolyJet is unique and virtually opposite to that of SLA:

- In PolyJet, 3D objects are built by selective deposition of UV curable compositions and irradiation of the deposited material using an inexpensive flood UV lamp. This is a significant advantage of PolyJet technology over other technologies, since upon completion of the building process, all the building materials are cured, reducing the risk of user exposure to any uncured and potentially hazardous material.
- In contrast, for example, in SLA, 3D objects are built by the unselective spreading of a UV curable liquid building material and selective curing of parts of the liquid building material using a UV laser beam. This process results in an object immersed in a bath of uncured building material, posing greater risks of user exposure to uncured building material.
- In addition, PolyJet uses at least two materials in the building process while SLA uses only one. This enables a smoother object surface and easy Support removal.

PolyJet enables building 3D objects directly from CAD data, in a swift, simple, cost effective and safe process, which does not necessitate tooling.

The objects may be built using a relatively wide range of materials, developed specially for PolyJet. PolyJet materials may be clear for applications requiring transparency or colored for better visualization. PolyJet materials may also be hard and tough, with a tensile strength of about 60 MPa and elongation to break around 20%. Elastic materials are also available. Elastic materials with tensile strength of about 1.5 MPa and elongation to break greater than 200% are already available. Despite this already relatively broad range of Modeling materials, one of the biggest challenges in further technology penetration is the development of improved Modeling materials for the most demanding applications.

POLYJET MATERIALS

UV curable Inkjet compositions formulated for use in two-dimensional (2D) Inkjet printers are typically formulated differently than compositions for PolyJet, for two types of reasons: Those related to differences between existing 2D and 3D printers and those related to the different requirements for the solid material final property in 2D and 3D printing.

As an example, 2D printers are usually designed to allow the use of only very low viscosity inks whereas PolyJet systems are designed to allow the use of higher material viscosities.

Regarding the properties required of the solidified materials in 2D printing, the most important properties are generally related to thin material layers, such as color, adhesion to the substrate, lightfastness, and scratch resistance, whereas in 3D printing, the most important properties are generally related to the bulk material mechanical and thermomechanical properties, for example tensile and flexural properties, impact resistance, and Glass Transition Temperature (Tg).

In addition, present PolyJet systems involve the use of two different building materials — a first material used mainly to build the 3D object and known as the Modeling material, and a second material, which is used temporarily during the building process and known as the Supporting material. Upon completion of the 3D object building process, the Supporting material is removed to leave the desired 3D object. Modeling and Supporting materials are both liquid reactive compositions which solidify by UV polymerization.

The Modeling material is a composition suitable for building a 3D object which, when cured, provides a solid material with mechanical properties that enables object building and handling.

The building materials comprise reactive components having polymerizable functional groups, initiator, surface-active agents and stabilizers.

The compositions are liquid at room temperature and have a viscosity lower than 20 cps at printing temperature, which is higher than room temperature. The curable components are generally combinations of (meth)acrylic monomers and oligomers.

An acrylic monomer is a low molecular weight functional acrylated molecule which may be, for example, esters of acrylic acid and methacrylic acid. Monomers may be monofunctional or multifunctional (for example, di-, tri-, tetra-functional).

An acrylic oligomer is a higher molecular weight functional acrylated molecule which may be, for example, polyesters of acrylic acid and methacrylic acid. Other examples of acrylic oligomers are the classes of urethane acrylates and urethane methacrylates. Urethane acrylates are manufactured from aliphatic or aromatic or cycloaliphatic diisocyanates or polyisocyanates and hydroxyl-containing acrylic acid esters.

The free radical photoinitiator may be any compound that produces a free radical on exposure to radiation, such as ultraviolet or visible radiation, and thereby initiates a polymerization reaction.

In Inkjet technology used by PolyJet for material deposition, the material liquid physical properties are of utmost importance. One of these properties is the material viscosity, which is of primary importance during droplet formation. The viscosity also influences drop velocity; the higher the viscosity, the slower the drop velocity and the smaller its volume.

Another important property is the building material surface tension. The surface tension has three main effects:

- **Drop formation**: The ability of the inkjet nozzle to form a droplet, as well as its volume and its velocity, are affected by the surface tension of the material. Within a certain range, the lower the surface tension, the smaller the volume of the droplet.
- **Drop spreading**: The PolyJet process is based on the formation of liquid films, rather than single drops, during the brief time interval between the drop's contact with a surface and the moment of UV exposure. Therefore, the surface tension (and the viscosity) is of major importance. The surface tension affects the magnitude of

drop spread and its uniformity during this time lapse. The time lapse affects the resultant drop spread and therefore the entire building process.

- **Interaction between liquids**: Since Modeling and Supporting compositions are in direct contact during the building process, prior to their exposure to UV radiation, the interfacial surface tension between these two liquids affects the building process. This interaction is the basis of the whole building process and in some cases is expressed in the 3D object surface smoothness, after the removal of Supporting material.

NEW DEVELOPMENTS IN INKJET 3D PRINTING

Although numerous improvements and different SSF approaches have been developed over the years, until very recently no systems enabling the building of objects using more than one Modeling material at a time had been developed. Only recently, during November 2007, Objet Geometries Ltd. from Israel disclosed a new technology breakthrough: *PolyJet Matrix*[6] and its new system: The Connex500.

This new technology enables the simultaneous use of two UV curable Modeling materials for building complicated assemblies and structures. In addition, this new technology offers the possibility of building Composite Materials or *Digital Materials*, as they are called by Objet Geometries Ltd.

DIGITAL MATERIALS

Digital Material is a new and special type of Composite Material, produced during the printing process by the selective deposition of different UV curable materials, from different inkjet nozzles, and according to a predefined Composite Material phase structure. This phase structure is design by Computer-Aided Design (CAD) software.

Because PolyJet Matrix technology allows the design and building of almost unlimited combinations of two UV curable materials, the number of *Digital Material* properties which can be achieved is also almost unlimited. For example, if one of the component of UV curable materials has a white color and the other a black color, *Digital Materials* with different gray tones can be produced during a single building process. A bicycle model where different parts were built simultaneously using several different *Digital Materials* is shown in Fig. 3.

Alternatively, if one of the component UV curable materials is soft and elastic and the other is hard and brittle, *Digital Materials* with intermediate or varying mechanical properties can also be produced. In addition, because *Digital Materials* are not simply a homogeneous combination of two different materials, but Composite Materials, where each component material maintains its own properties within a microscopic phase range, their properties are not just the average of the properties of the component materials, but much more. Because *Digital Materials* are defined by means of precise software

Fig. 3. Inkjet 3D model of a bicycle. The model was built by PolyJet Matrix technology, in a single printing process by the simultaneous use of several *Digital Materials*.

design, special anisotropic materials and graded materials are also possible.

REFERENCES

1. U.S. Patent No. 6,569,373
2. U.S. Patent No. 4,575,330
3. U.S. Patent No. 5,733,497
4. U.S. Patent No. 5,387,380
5. U.S. Patent No. 5,855,836
6. U.S. Patent No. 7,300,619

CHAPTER 14

Printing Bioinks with Technologically Relevant Applications

Leila F. Deravi, David W. Wright, and Jan L. Sumerel*

Department of Chemistry, Vanderbilt University Nashville TN, 37235

Fujifilm Dimatix, 2230 Martin Ave, Santa Clara, CA, 95050

INTRODUCTION

For decades, researchers have been searching for alternative methods to pacify the rate of waste production, the cost of spending, and the energy intensive techniques commonly associated with nanofabrication and semiconductor facilities. The cost associated with supporting laborers and instruments, alone, has been estimated to reach an outstanding $100 billion dollars per fabrication facility by the year 2020, as the demand for smaller, lighter, and faster materials continues to grow.[1,2] As processing techniques continue to move from the R&D sector to many industrial and laboratory settings, researchers have been searching for alternative methods that would eliminate the cost of spending and reduce the rate of waste production inherent in nanofabrication facilities. This drive has harnessed alternative approaches, utilizing water soluble, biological materials, or *biomaterials*.

It is believed that patterning biomaterials has the potential to produce functional, reactive materials under ambient conditions. From manual stamping to computer-based inkjet printing, the availability of current materials deposition techniques has offered unique approaches for immobilizing biomaterials with particular applications. Specifically, the controlled deposition of proteins, cells, polymers, and antibodies required for cell-based biosensors,[3–7] bone tissue engineering supports,[8] or antimicrobial devices,[9] has gained attention as the demand for hybrid biological devices has increased.[10,11] In this chapter, we will present a brief overview of both contact and non-contact printing techniques that have been used to deposit biomaterials with applications ranging from microscale medical devices to 3D tissue engineering platforms in living systems.

MICROCONTACT PRINTING

Microcontact printing (μCP) is one direct contact technique that has been optimized to print biomaterials. Some patterned materials include printed lipids or polymers into tissue culture flasks or printing long-chain alkanethiols for organic-based sensors onto gold substrates.[12,13] Prior to each printing process, an elastomeric polydimethylsiloxane (PDMS) stamp is molded against a preformed master with defined patterns for a specific application; the master is often generated using photolithography.[12] Once the PDMS stamp is ready for use, it is traditionally immersed in a composite solution, or *ink*, allowed to dry over a period of time, and stamped onto a substrate, often leaving a patterned monolayer with nanoscale resolution.[12] The degree, or amount, of stamping is relative to the pressure applied to the stamp and its contact time (10–1000 s) on a substrate.[13]

Recently, Zhao and coworkers have utilized a modified version of this technique to pattern lipid tubules onto gold substrates.[14] In these experiments, a PDMS stamp was created having parallel, recessed channel walls (4–7 μm apart, 0.8 μm high, and 1.0 μm, wide). The stamp was placed directly onto a glass substrate,

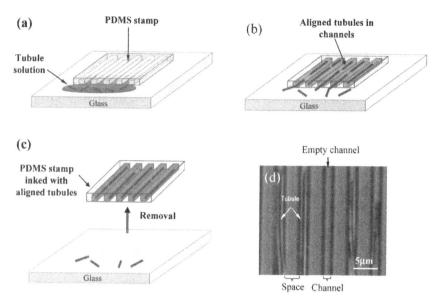

Fig. 1. Illustration of microcapillary capabilities of PDMS stamp, where a composite suspension of lipid tubules was drop casted onto a glass substrate (a–c). Once the lipid tubules were assembled into the PDMS channels, their confinement was confirmed through transmission mode optical microscopy. Reprinted with permission from ref. 14. Copyright 2008 American Chemical Society.

where a solution of 1,2-bis(tricosa-10,12-dinoyl)-*sn*-glycero-3-phosphocholine ($DC_{8,9}PC$) lipid tubules was drop casted along its open ends (Fig. 1). The solution was then aligned into highly ordered two-dimensional (2D) arrays, where they settled into the microfluidic channel walls of the PDMS stamp. The lipids were later repatterned through a secondary stamping process (2 hour contact time onto gold coated mica substrates), producing a stable, three-dimensional (3D) lipid pattern that was roughly 360 nm tall. Integrating such 2D ordered patterns into 3D junctions is believed to potentially redefine new types of biomaterial microdevices.[14]

ELECTRON BEAM RADIATION

Electron beam (e-beam) irradiation of polymeric materials into specific micropatterns has been used to investigate the adhesive nature of a variety cells. E-beam radiation is a soft lithography technique incorporating a metal mask that can withstand the irradiation, using an Area Beam Processing System. Successful cell micropatterning could not only be used to document the binding properties of a specific cell on a substrate, but it could also be used to identify its primary binding sites.[15] Furthermore, seeding cells onto controllable, dynamic substrates can also expose a number of interesting properties necessary for important applications, such as cell-based biosensors or even defining basic cellular mechanisms.[15] In order to realize these applications, a number of parameters have been tested.[15] Factors used to control cell adhesion at the molecular level, such as the effect of external stimuli, the polarity of the patterned substrate, or the effect of surface selective adhesion, can be varied to produce a number of outstanding properties.[15]

Both cell adhesive and non-cell adhesive materials can be patterned onto a surface to eliminate the effects of non-specific cell binding. Common biological components that do not bind to cells, such as polyethylene glycol (PEG), bovine serum albumin (BSA), or amphiphilic phospholipids, can be patterned to resist the non-specific protein adsorption that is associated with cell media or protein byproducts secreted from cells. On the other hand, extracellular matrix (ECM) glycoproteins, such as collagen and fibronectin, contain a specific, linear peptide sequence containing arginine, glycine, and aspartate (RGD) that have been shown to interact with cell membrane proteins for specific cell binding.[15] During these printing cycles, a predefined mask is irradiated on the surface of a substrate immersed in a solution containing one of these materials, dried in air, and removed, leaving a specific micropattern.

Okano and coworkers have patterned poly(N-isopropylacrylamide) (PNIPAAm), a material similar to BSA and PEG in that it does not adhere to cells at room temperature.[16,17] In this experiment, a composite solution of PNIPAAm dissolved in propanol (55 wt%) was uniformly coated inside a commercial cell culture dish. The polymer was then patterned, using standard e-beam lithography, onto the surface of a cell culture dish to test the dynamic behavior of cells for potential clinical applications. A metal mask (60 mm o.d.

with 1 mm circular holes) was then placed into the dishes and subjected to e-beam irradiation (0.25 MGy).[17] The PNIPAAm-grafted cell culture dishes were washed extensively, and their properties were studied. Interestingly, PNIPAAm alters its solubility and subsequent hydrophilicity at 32°C, its lower critical solution temperature (LCST), making it a viable template for controlling the reactivity of cell patterns.[16] When cells were seeded below the LCST, they adhered only to the regions that were not patterned with PNIPAAm. Once the temperature was raised above the LCST, the cells adhered to all surfaces, providing an interesting example of region selectivity of cells.[17]

NON-CONTACT PRINTING

Although contact printing techniques have been used to create intricate patterns at relatively low costs, the arrays are often plagued by constant stamp fatigue and unstable patterns; in fact, some nanopatterned surfaces have been reported to experience post-deposition degradation.[10,18] More importantly, because contact printing requires manual deposition, it is not suitable for depositing production scale quantities necessary to compete with current high-throughput facilities. Techniques that incorporate robotic deposition, such as direct ink write (DIW), syringe solenoid jet printing (SSP), or inkjet printing (IJP), are some examples of alternative, non-contact deposition techniques that have been developed to alleviate the element of user error for printing biomaterials.[19–21]

In contrast to other prototyping techniques, non-contact printing is a relatively straightforward fabrication process due to its computer graphical interface. In general, many formats of two-dimensional drawings, pictures and structures can be converted to a bit map image, which can then be interpolated into X- and Y-coordinates to deposit materials in a printed pattern. Most non-contact printers also rely on the chemical properties, specifically the surface tension and viscosity of the material being printed. A composite solution, or *ink*, is tailored specifically to the printing device that is employed, and the pattern resolution (drops per inch (dpi)) is affected by the method and properties of a print cycle. The thickness and dimensions of a patterned film are not only related to the properties of the *ink* but are also dependent on a number of controllable parameters that include but are not limited to, nozzle diameter, solvent, and substrate.[22] Once released from a syringe or nozzle, the droplet maintains a spherical shape, with dimensions of πr^3; however, once it hits the surface of a substrate, it will adjust its geometry over post-impact time to form one of the three primary geometries: a dome, a cylinder or a torus. Such high resolution patterning techniques have many potential biological and medical applications that will be discussed below.

Continuous Flow Printing

DIW and a SSP rely on the *direct writing* of continuous flow ink filaments through a cylindrical nozzle (DIW) or syringe (SSP). DIW has been used to deposit a number of composite inks, including organic ligands, polyelectrolyte composites, or colloidal solutions, and the resultant patterns are developed through layer-by-layer deposition.[23] Xu and coworkers have used DIW, for instance, to pattern ($40\,\mu ms^{-1}$) polyamine rich inks that were mineralized to yield microscale feature sizes.[20,23] In these experiments, the polyelectrolyte ink is loaded into a syringe and robo-casted onto a substrate using computer animated design. The patterned surfaces were then mineralized through hydrolysis with silicic acid to condense micropatterned silica, similar to micropatterns found on the cell walls for marine diatoms.[20] Auger electron microscopy was used to map the distribution of silica along the patterns, and it was confirmed that both silicon and oxygen were distributed uniformly about the pattern.

SSP has the capability for rapid actuation, making the solenoid type of dispenser ideal for non-stop reagent dispensing, as they have become integral components to high-throughput laboratory applications in pharmaceutical industries. These printers incorporate a microsolenoid valve and a syringe pump, similar to the nozzle used in DIW that is used to compress the fluid in the reservoir. When opened, the solenoid valve creates a pressure wave that forces fluid through the orifice.[24] Currently, there are three methods of solenoid dispensing: aspirate-dispense, flow-through, and isolated. Once expelled from a syringe, droplets are released in a continuous flow, which yield large, non-uniform patterns. Although these patterns do not possess the resolution necessary for developing 3D biomaterial devices, they can be used for liquid dispensing on larger scales.[19,24]

Drop-on-Demand Printing

The pressing need for smaller drop sizes, faster printing speeds, higher accuracy in drop placement, and higher resolution patterns with biological applications has inspired the immergence of drop-on-demand printing, or inkjet printing, which has been adapted for biomedical materials applications under ambient reaction conditions. *Just Push Print* — a command, specific to operating systems in almost every home or office can now be applied to the direct immobilization of functional three-dimensional (3D) devices made from biomaterials. In contrast to other prototyping methods, inkjet printing is a relatively straightforward fabrication process due to its user-friendly computer graphical interface.

Derived from standard desktop printers, drop-on-demand techniques, such as electrohydrodynamic jet (e-jet) printing, solid free-form fabrication (SFF), or piezoelectric inkjet printing, are among a few notable examples of rapid prototyping techniques that have been developed to pattern biomaterials.[22,25,26] With the incorporation of a user friendly computer interface, these techniques have been employed to process a number of multiplexed, biomaterial constructs without the use of masks, stamps, or any other time consuming processing equipment. The design of such constructs with minimal feature sizes in the microliter to picoliter resolution has been demonstrated and will be discussed below.

Inkjet Printing of Biomolecules

The print head of standard, color inkjet printers (Cannon, Hewlett-Packard) is composed of a nozzle, heater, manifold chamber, and restrictor.[27] Once the resistive element is heated (300°C) the fluid ink begins to bubble, eventually releasing a droplet that flows from the restrictor into the chamber of the print head. Such high resolution (300 dpi) inkjet print heads have been optimized to print a range of colored *inks* by sequestering each color from others in the cartridge.[28] Goldmann and coworkers have taken advantage of such properties to pattern complex, DNA microarrays onto solid substrates for hybridization experiments.[28]

In these experiments, DNA was printed, where the plasmid (600 bp fragment of mouse glyceraldehyde 3-phosphate dehydrogenase (GAPDH) in Bluescript II KS$^+$ (BS)) was diluted ($2 \mu g \mu L^{-1}$) in equal parts of water and regular blue ink (Pelikan 4001). The composite ink was allowed to incubate for 5 min at room temperature and 10 min on ice before 4 M NH$_4$-acetate was added to neutralize the ink. The denatured DNA (2–3 mL) was loaded into an empty Hewlett-Packard ink cartridge (HP DJ 500C inkjet printer) and was printed directly onto nitrocellulose or nylon membranes using a Microsoft Write software.[28] Once printed, the arrays were incubated at 80°C for 90 min and stored at 4°C until hybridization.

Printed membranes were hybridized with a purified 600 bp GAPDH insert and labeled with radioactive probes. Visible signal was detected after 5 min of exposure with a digoxigenated probe followed by 30 min incubation with the radiolabeled probe. Mock hybridization experiments verified that there was no observed, unspecific signal, confirming that inkjet printing using standard desktop printers can be used to directly transfer DNA onto membranes for further reactivity.

E-Jet Printing

Modified versions of desktop printers have recently been developed to deposit a wider range of materials without exposing them to high temperature environments. One such example is a drop-on-demand technique, electrohydrodynamic jet (e-jet) printing, that utilizes an internal electric field to pattern submicrometer range droplets onto a conducting surface.[26] In this process, a syringe pump or a pneumatic pressure controller is connected to a glass capillary (i.d. $= 0.3$–$30\,\mu$m, o.d. $= 0.5$–$45\,\mu$m), which serves as a single printing nozzle. To minimize the probability of particle clogging, the capillary interior is coated with a thin layer of gold. It is also supported by a fixture, which is connected to a pump as the print head. A d.c. voltage is applied between the nozzle and a conducting substrate inducing electrohydrodynamic flow. Once a critical electric field is reached, the electrostatic (Maxwell) stress of the fluid ink overcomes its capillary tension, and a droplet is released onto the grounded substrate.

This process has been utilized to deposit a range of electronic devices of printable organic and inorganic inks onto a variety of substrates with $\sim$10 μm feature sizes.[26] Park and coworkers have printed thin-film transistors (TFTs) onto plastic substrates. In these experiments, single-wall carbon nanotubes (SWNTs) were printed as the semiconducting material and were characterized. Initially, the SWNTs were suspended in a solution of octyl-phenoxy-polyethoxyethanol (6.9 mg ml^{-1}) as the ink and were printed onto glass and quartz substrates. The arrays were shown to consist of 2.5 SWNTs/10 μm spot sizes. The current output of the subsequent patterns were shown to increase linearly with $1/L$, where the *on* to *off* ratio currents ranged from $\sim$1.5 and 4.5, providing an excellent example of printable electronics. In this example, high resolution e-jet printing has been demonstrated as a viable deposition technique that produced reactive, high resolution patterns; however, applications of e-jet printing are still limited because of the substantial charge associated with the deposited droplets inherent in the printing process.

SFF Printing

Solid free-form (SFF) printing has been used to design complex scaffolds for organ printing or tissue engineering.[20] This technique builds 3D structures using layer-by-layer deposition. Some SFF systems are commercially available and have been shown to process inks using either a chemically or thermally powered nozzle. In this process, the liquid monomer or sintered powder is photopolymerized for printing. Organ printing has recently gained attention,[29] partially due to organ transplant shortages. An organ can be described as an anatomical description of a group of cells that function together in eukaryotic organisms; they are maintained by the organism's nutrient supply and can be described as multicellular *devices* that appear as simple looking as a leaf on a pea plant or as complex as a multifaceted mammalian heart. While essential parameters required for successful organ printing are still unknown, inkjet printing using SFF can still be utilized as a viable method with controllable parameters for depositing single components of organs.

It is understood that the vasculature density of primary organs is one of the most important components for adequate organ function; without which, essential processes such as cell apoptosis or necrosis would be inhibited.[29] In their experiments, Mironov and coworkers viewed the embryonic organ tissues as standard viscoelastic fluids with specific flow and fusion kinetics. For this reason, a specific ink was synthesized for printing, so that once deposited, the cells would fuse together to form a disc for tubular tissue.

To begin the printing process, 3D image files of the living organs are transposed into 3D computer-aided design (CAD) file that produces a 3D computer aided manufacturing (CAM) file. The printer's software then registers the CAM file and accurately deposits eukaryotic cells appropriate to that organ type in a layer by layer construction method. The final product is a 3D structure that resembles the starting imaged organ in both size and cellular componentry. For proper organ printing to be realized, basic theories of developmental biology must hold true post-processing, as well.

Once the ink is synthesized, patterns are created using a computer graphical image and layer-by-layer deposition of a thermoreversible gel for cell fusion (Fig. 2). Hydrogels were used to mimic of the natural habitat of cells, the ECM, which provides an environment for the cells to attach and grow, permitting fusion of cell aggregates. In their experiments, Mironov and coworkers printed multicellular spheroids composed of Chinese Hamster Ovary cells into a collagen hydrogel (1 mg ml^{-1}).[29] This dynamic environment is just one example of how the precise placement of cells or proteins

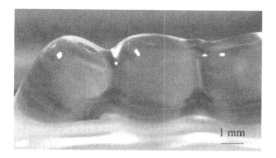

Fig. 2. Printed alginic acid (2% in 0.25M $CaCl_2$ buffer) tubes. $CaCl_2$ is known to promote cross-linking of alginate, producing a resultant 3D construct. Picture was taken immediately after printing.[20] Copyright Wiley-VCH Verlag GmbH & Co. KGaA. Reproduced with permission.

within a hydrogel using inkjet printing can be used as a scaffold for tissue engineering.

Piezoelectric Inkjet Printing

Piezoelectric inkjet deposition is an alternative to e-jet and SFF printing, as it is a non-contact, non-destructive technique for patterning biomaterials ranging from multicellular components to electronic devices. The Dimatix Materials inkjet Printer (DMP), specifically, uses a piezoelectric print head that consists of a piezoelectric transducer, 16 nozzles (21.5 μm diameter, spaced 256 μm apart), a manifold, an ink pumping chamber, and a fluid inlet passage. A user-controlled voltage is applied to the print head that induces a deformation of the lead zirconate titanate (PZT) piezoelectric transducer.[22] This deformation creates mechanical vibrations that generate acoustic waves, releasing a droplet from the chamber through the nozzle. Other piezoelectric print heads are categorized based on the deformation mode of the transducer (e.g., squeeze mode, bend mode, push mode, or shear mode) powered by voltage to a thin piezoelectric ceramic structure, which is constructed in the plane of the wafer.

Although it is still a relatively new deposition technique, the DMP has already been recognized as a highly reproducible rapid prototyping technique with nanoscale resolution.[22] Materials such as monofunctional acrylate esters for the advancement for medical prostheses, sinapinic acid for more sensitive MALDI-TOF matrices, and hybrid composite multiwalled carbon nanotubes (mwCNT) for direct electrochemical oxidation, have already been printed, demonstrating the versatility of the printer.[30] In the case of mwCNT prining, a mwCNT composite solution (1 mg ml^{-1}) was synthesized in the presence of propylene glycol, salmon sperm DNA (1 mg ml^{-1}), and 1 μg ml^{-1} of a DNA probe (DAPI). Once synthesized, the solution was inserted into a standard print cartridge (1.5 mL), and printed onto gold, glass, and sapphire substrates at variable voltages. The average spot size of the mwCNT averaged to be 98.4 $\pm$ 5.4 μm, where fluorescence was localized uniformly across the surface area of each spot.

CONCLUSION

Regardless of the system, the techniques available for printing biomaterials have expanded to produce highly functional materials. Contact printing techniques, such as μCP or e-beam lithography, produce patterned monolayers or grafted polymers with nanoscale resolution. Non-contact deposition, such as piezoelectric, e-jet, or SFF printing, are rapid prototyping techniques that pattern a range of materials onto moving platforms. By controlling the composition of the printed material, the specific pattern, and the number of print cycles, these highly reactive 2D and 3D constructs could serve as the basis for future functionalized surfaces.

REFERENCES

1. Wolf S. (2004) *Microchip Manufacturing*. Lattice Press: Sunset Beach.
2. Kumar U, Shete A, Harle AS, Kasyutich O, Schwarzacher W, Pundle A, Poddar P. (2008) Extracellular bacterial synthesis of protein functionalized ferromagnetic Co_3O_4 nanocrystals and imaging of

self-organization of bacterial cells under stress after exposure to metal ions. *Chem Mater* **20**: 1484–1491.

3. Pudas M, Hagberg J, Leppavuori S. (2004) Printing parameters and ink components affecting ultra-fine-gravure-offset printing for electronics applications. *J Eur Ceram Soc* **24**: 2943–2950.

4. Ye Y, Ju H. (2005) *Biosens Bioelectron* **21**: 735–741.

5. Coe S, Woo WK, Bawendi M, Bulovi V. (2002) *Nature* **420**: 800–803.

6. Radsak MP, Hilf N, Singh-Jasuja H, Braedel S, Brossart P, Rammensee H-G, Schild H. (2003) *Blood* **101(7)**: 2810–2815.

7. Clokey GV, Jacobson LA. (1986) *Mech Ageing Dev* **35**: 79–84.

8. Coleman JN, Khan K, Blau WJ, Gun'ko YK. (2006) *Carbon* **44**: 1624–1652.

9. Ryu B-H, Choi Y, Park H-S, Byun J-H, Kong K, Lee J-O, Chang H. (2005) Synthesis of highly concentrated silver nanosol and its application to inkjet printing. *Colloids Surf A Physicochem Eng Asp* **270–271**: 345–351.

10. Xia Y, Rogers JA, Paul KE, Whitesides GM. (1999) Unconventional methods for fabricating and patterning nanostructures. *Chem Rev* **99**: 1823–1848.

11. Dong H. Carr WW. (2006) An experimental study of drop-on-demand drop formation. *Phys Fluids* **18**: 72102.

12. Mirksich M, Whitesides GM. (1995) Patterning self-assembled mono-layers using microcontact printing: a new technology for biosensors? *TIBTECH* **13**: 228–235.

13. Synder PW, Johannes MS, Vogen BN, Clark RL, Toone EJ. (2007) Bio-catalytic microcontact printing. *J Org Chem* **72**: 7459–7461.

14. Zhao Y, Fang J. (2008) Direct printing of self-assembled lipid tubules on substrates. *Langmuir* **24**: 5113–5117.

15. Nakanishi J, Takarada T, Yamaguchi K, Maeda M. (2008) Recent advances in cell micropatterning techniques for bioanalytical and biomedical sciences. *Anal Sci* **24**: 67–71.

16. Tsuda Y, Kikuchi A, Yamato M, Nakao A, Sakurai Y, Umezu M, Okano T. (2005) *Biomaterials* **26**: 1885.

17. Yamato M, Konno C, Utsumi M, Kikuchi A, Okano T. (2002) *Biomaterials* **23**: 561.

18. Nelson CM, Raghavan S, Tan JL, Chen CS. (2002) Degradation of micropatterned surfaces by cell dependent and independent process. *Langmuir* **19**: 1493–1499.

19. Coffman EA, Melechko AV, Allison DP, Simpson ML, Doktycz MJ. (2004) *Langmuir* **20**: 8431–8436.

20. Boland T, Xu T, Damon BJ, Cui X. (2006) Application of inkjet printing to tissue engineering. *Biotechnol J* **1**: 910-917.

21. Kisailus D, Truong Q, Amemiya Y, Weaver JC, Morse DE. (2006) Self-assembled bifunctional surface mimics an enzymatic and templating protein for synthesis of a metal oxide semiconductor. *Proc Natl Acad Sci USA* **103**: 5652–5657.

22. Deravi LF, Sumerel JL, Gerdon AE, Cliffel DE, Wright DW. (2007) Output analysis of materials inkjet printer. *Appl Phys Lett* **91**: 113114–113116.

23. Lewis JA, Gratson GM. (2004) Direct writing in three dimensions. *Mater Today* July/August: 32–40.

24. Espina V, Mehta AI, Winters ME, Calvert V, Wulfkuhle J, Petricoin III EF. Liotta LA. (2003) Protein microarrays: Molecular profiling technologies for clinical specimens. *Proteomics* **3**: 2091–2100.

25. Boland T, Tao X, Damon BJ, Manley B, Kesari P, Jalota S, Bhaduri S. (2007) Drop-on-demand printing of cells and materials for designer tissue constructs. *Mater Sci Eng C* **27**: 372–376.

26. Park J-U, Hardy M, Kang SJ, Barton K, Adair K, Mukhopadhay DK, Lee CY, Strano MS, Alleyne AG, Georgiadis JG, Ferreira PM, Rogers JA. (2007) High-resolution electrohydrodynamic jet printing. *Nat Mater* **6**: 782–789.

27. Bae KD, Baek SS, Lim HT, Kuk K, Ro KC. (2005) Development of the new thermal inkjet head on SOI wafer. *Microelectronic Eng* **78079**: 158–163.

28. Goldmann T, Gonzalez JS. (2000) DNA-printing: utilization of a standard inkjet printer for the transfer of nucleic acids to solid supports. *J Biochem Biophys Methods* **42**: 105–110.

29. Mironov V, Prestwich G, Forgacs G. (2007) Bioprinting living structures. *J Mater Chem* **17**: 2054–2060.

30. Sumerel JL, Lews J, Doraiswamy A, Deravi LF, Sewell SL, Gerdon AE, Wright DW, Narayan RJ. (2006) Piezoelectric inkjet processing of materials for medical and biological applications. *Biotechnol J* **1**: 976–987.

CHAPTER 15

Printed Electronics

Vivek Subramanian
Department of Electrical Engineering and Computer Sciences
University of California, Berkeley

INTRODUCTION

Over the last decade, printed electronics has received substantial attention as a potential application of inkjet technology. Conceptually, the goal is to use printing technology as a replacement for conventional photolithography-based semiconductor manufacturing. This is expected to result in a substantial cost reduction for the realization of simple semiconductor systems on cheap, flexible substrates such as plastic, steel foils, etc.

Within this chapter, the conceptual basis for printed electronics is reviewed, including a discussion of the motivation for the same technology, as well as an analysis of the potential applications for printed electronics. Next, classes of electronic materials that are suitable for inkjet printing are reviewed, and finally, methods of fabricating printed electronic devices and circuits are discussed. Based on this discussion, the outlook for printed electronics is summarized.

Motivation for Printed Electronics

Fundamentally, the global interest in printed electronics is motivated by the fact that printing in general, and inkjet printing in particular, represents a potentially low cost means of depositing different electronic materials on a substrate in a spatially-specific manner. This has particularly important implications on the cost of realizing electronic systems.

Conventional electronic systems are fabricated using a subtractive photolithographic process.[1] To create a pattern in an arbitrary electronic material (for example, a conductor, dielectric, or semiconductor), the process proceeds as follows — (1) the electronic is deposited in a blanket layer, typically using a vacuum-based process such as chemical vapor deposition or physical vapor deposition; (2) the substrate is then coated with photoresist; (3) the photoresist is exposed using a photolithographic system through a mask; (4) the photoresist is developed to remove the resist from mask-determined regions of the substrate; (5) the deposited film is etched using a wet chemical or dry (i.e., plasma-based) etch; the deposited film is only etched from regions where the photoresist has previously been developed away; (6) the remaining resist is stripped away to expose the patterned film, and finally, (7) the substrate is cleaned to remove residue from the resist and etch process. This seven-step process is repeated multiple times with multiple masking and etch steps to create sophisticated devices and circuits.

The above technique is called subtractive since patterns are created by "removing" material, i.e., blanket films are deposited and subsequently patterned. The consequence of this subtractive process is that numerous steps are required to create a single patterned film. This increases complexity of the process, and also increases capital expenditure, since appropriate tooling is required to realize these complex process flows.

Printing in general, and inkjet printing in particular, potentially allows the realization of an alternate manufacturing paradigm, namely additive processing. In additive processing, patterned deposition is used to deposit the film directly. In other words, the overall process has conceptually reduced from seven-steps per patterned layer to a single step. Inkjet printing is an excellent example of an additive printing technique; inks are printed only at the locations where they are desired.

Based on this step-by-step analysis, it is possible to postulate on the advantages and disadvantages of printed electronics over conventional microelectronic processes. Since printed electronics can potentially reduce overall process complexity, it is expected to realize process flows with reduced capital expenditure, and excellent compatibility with low cost, large area substrates such as plastic, foils, etc. Therefore, it is expected that the cost per unit area of printed electronics will be one to three orders of magnitude less than the cost per unit area of photolithography-based subtractive processing. On the other hand, conventional photolithography is capable of realizing features with linewidths in the 50 nm range in conventional microelectronics, and in the range of 1–3 microns in large area displays. State-of-the-art industrial inkjet printer can only realize feature sizes in the range of 20 microns, and even in research, pushing inkjet technology below 1 micron is extremely challenging. Therefore, while the cost per unit area of printed electronics is expected to be much lower than the cost per unit area of subtractive silicon processing, the cost per transistor in conventional microelectronics is much lower than the cost per transistor in printed electronics, simply because the size of a transistor is so much smaller.

Therefore, it is possible to summarize the motivation for printed electronics quite simply. Printed electronics is attractive as a means of fabricating electronic systems where reduced cost per unit area is an advantage. Based on the poor linewidth of printed electronics, this will almost certainly only be true for systems with low functional density, i.e., systems with no need for dense arrays of tiny transistors, etc. Printed electronics also potentially enables the use of low cost, flexible substrates such as plastics and metallic foils, and additionally, by using different inks in an inkjet printer, it also allows for the easy deposition of a range of materials onto a substrate in a spatially specific way.

Applications of Printed Electronics

Given the constraints of low cost per unit area, but relatively high cost per function of printed electronics, various target applications are being pursued. All the major applications are characterized by needing large area operation and/or compatibility with flexible substrates. The applications that are generally considered the most promising are (1) flexible displays, (2) RFID tags, and (3) sensors.

Flexible Displays

The application of printing to flexible displays makes immediate intuitive sense. Displays benefit from flexibility for form factor and manufacturing reasons. The idea of a roll-up display is attractive for a broad range of consumer appliances.[2] Such displays can be used to reduce the size of the display device when stored, while simultaneously improving weight and mechanical reliability by eliminating the use of brittle, crack-prone glass. Printing is potentially interesting for realizing both the display elements themselves, as well as for the circuitry driving the display pixels in an active matrix application.

Currently, various display elements are being considered for flexible displays, including liquid crystal displays, organic light emitting diodes, and electrophoretic displays. Liquid crystal displays make use of a light switch implemented using a liquid crystalline small organic molecule. By either inserting this molecule between polarizing plates or by implementing scattering based display elements, a display can be realized. Printing has been studied for use in liquid crystals, both for printing the display pixels themselves (albeit only in research), and also for printing color filters to achieve RGB color control.[3] The latter application has had some success, with several major companies working in the field. It is expected that products with inkjet printed color filters will enter the market over the next several years. In addition, many liquid crystal displays are implemented with a transistor to drive each pixel. This transistor does not require particularly high resolution, and is potentially printable, as will be discussed below.

Organic light emitting diodes are currently being developed by numerous companies and research institutions worldwide. From the perspective of printed electronics, polymer-based light emitting diodes are very attractive. Polymer LEDs are essentially diodes

Color Filter (can be printed)

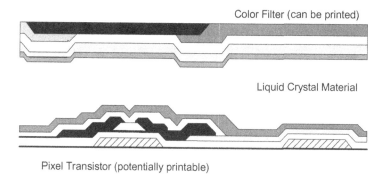

Liquid Crystal Material

Pixel Transistor (potentially printable)

Fig. 1. Cross-section of a typical active matrix liquid crystal STN pixel.

with light-emitting polymer sandwiches.[4] As current flows through the diode, electrons and holes recombine to emit light. Various inkjet printable polymer light emitting materials are commercially available, and several companies and research institutions are attempting to develop products based on the same technology using a variety of printing techniques, including inkjet printing.[5] The advantage of the PLED over the LCD is that it is a much simpler structure and is thus potentially much more amenable to printing. On the other hand, unlike the LCD, which does not generate its own light but is simply a light switch for an external light source, the PLED actually generates its own light, and therefore requires more energy to operate, in the form of fairly high operating currents. As a result, it is unlikely that such displays will be powered by printed transistors in the near future, since the performance requirements for many such display applications will be out of the reach of early printed transistors.

The final display element that has received substantial attention in recent years as a good application for printed electronics is the electrophoretic display (often called e-paper).[6] The most common electrophoretic display element is a bistable element consisting of a dye-filled cell containing charged colored particles of Titania. When an external voltage is applied across the cell, the particles move up or down, depending on the polarity of the voltage, to make the cell appear either light or dark. Since the cell does not emit light, it

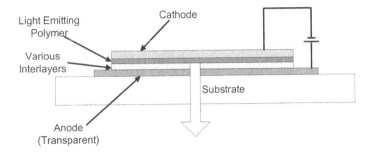

Fig. 2.   Cross-section of a conceptual polymer light emitting diode (PLED).

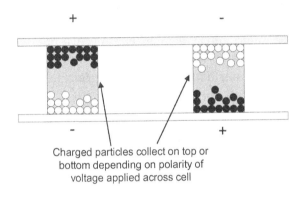

Fig. 3.   Cross-section of a conceptual electrophoretic cell.

does not require high current driving. This makes it potentially very compatible with printed transistors, and, indeed, there are several companies currently working on developing printed displays based on electrophoretic cells.[7,8] While the cell itself is not printed, the intended application, i.e., as an electronic paper or e-book with low speed, relatively low resolution images and text, makes it a very strong candidate for initial application of printed electronics.

In all three display technologies above, there are two separate considerations with regard to printing. The first consideration is whether the display element itself is printable. Here, clearly PLED has made the most significant progress, though arguably the printing

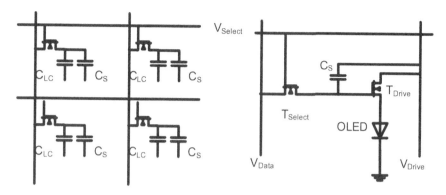

Fig. 4. (Left) Schematic representation of an active matrix LCD display, showing single transistors driving capacitive pixel elements. (Right) OLED displays, on the other hand, require current-based driving, and therefore, multi-transistor pixel architectures are more common.

of color filters for LCD is closest to manufacturing. The second consideration is whether the display element is compatible with a printed backplane of transistors. Here, electrophoretic displays are probably at the greatest advantage for two reasons. First, these cells are capacitive in nature, i.e., they do not require high drive currents to operate (as shall be shown later, this is a significant concern for printed electronics, at least in the near term). Second, the usage model for these displays is focused on e-paper applications that do not require high operating speed, so the poor performance of initial printed transistors is not expected to be a major limiter. After electrophoretic displays, LCD's are arguably the next most compatible in terms of performance requirements, for similar reasons. OLEDs, on the other hand, are not capacitive elements; rather, they are current-driven displays. As a result, they are less compatible with printed transistor performance specifications, though some demonstrations of polymer transistor-driven OLEDs have been shown. Note that OLEDs also typically require more complex pixel circuitry, which is also problematic given the poor linewidth of printing technology.

Summarizing the state-of-the-art and outlook for printed displays, it is clear that opportunities do exist for printing displays, particularly on the low end of performance. Within the next few years, opportunities for such systems will become clearer. A key driver for such displays is the ability to integrate transistors for active matrix display addressing. Thus, the next topic in this chapter will focus on applications that are more transistor-centric.

Printed RFID Tags

Printed RFID tags are a perfect example of an application where the cost advantages of printing are a primary driver for the development of the technology. RFID tags come in varying degrees of complexity, ranging from silicon-based solutions incorporating sophisticated microcontroller chips allowing encryption, bidirectional communication, multistate reprogrammability, etc., to simple RF barcodes that periodically and unintelligently repeat an identity bit string when power is available. It is the latter end of the RFID spectrum that is the focus of printed RFID, as will become apparent shortly.

Silicon-based RFID tags are widely available today, and are used in numerous applications ranging from asset management and inventory control to security and transit applications.[9] In its simplest form, an RFID tag consists of a digital finite state machine driving a modem connected to an antenna. The antenna is responsible for uni- or bi-directional communication with a reader, as well as for providing power to the tag.

The antenna in an RFID is typically implemented as a spiral inductor or as a dipole antenna, depending on the frequency of operation of the tag. This frequency of operation depends on the application, government-imposed standards, physical constraints, etc. The most common frequencies for operating RFID tags are <125 kHz (called the LF band), 13.56 MHz (called the HF band), ~900 MHz (called the UHF band), and 2.4 GHz (called the microwave band). For various reasons, 125 kHz tags are not compatible with planar processing and thus will not be considered here.

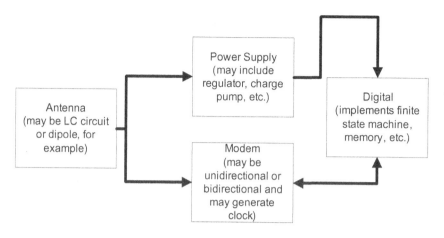

Fig. 5. Architectural overview of an RFID tag.

The antenna for the other frequencies is typically moderately large, covering an area of several cm square. As a result, in silicon-based RFID tags, the antenna is processed separately on a piece of plastic or paper, and then the silicon chip is mounted onto the antenna using an attachment process.

Examining a typical RFID tag, several points of note are apparent. The first point is that the size of the tag is dominated by the antenna, which is printed at a relatively low resolution; the size of the circuitry is a relatively small fraction of the overall tag size. The second point is that the cost scaling of silicon-based RFID is limited by the cost of attaching the tiny silicon chip to the antenna; thus, silicon-based RFID only benefits partially from cost reduction in the silicon microelectronics industry. The third point is that the transistor performance requirements for simple RF barcodes are not outrageous, and are potentially in the range of what is achievable using printed electronics.

These three points have resulted in substantial interest in printed RFID. Conceptually, the idea is to replace the silicon RF chip in RF barcodes with a printed RF circuit. This circuit may be printed at the same time as the antenna or, at the very least, may be printed on the strap and attached to the antenna using a low cost attachment technique. In contrast to silicon RFID, therefore, printed RFID is expected to result in low cost RFID tags, albeit only at the very low end of RFID, i.e., RF barcodes. For high-end applications, or applications requiring bidirectional communication, silicon RFID will still dominate. However, the postulated high volumes of RF barcode markets have driven substantial interest in printed RFID as a result.

A prototypical printed RFID tag will obviously be limited by the performance of the printed transistors. For this reason, 13.56 MHz is considered the most promising frequency for printed RFID. An archetypal 13.56 MHz RF barcode consists of an inductive antenna driving a rectifier for power. This power is used to drive a digital circuit generating a stream of information, which is in turn modulated back onto the inductor. The resulting change in loading is detected by the reader, thus enabling the reader to determine the stream of bits being generated by the tag. The data rate of this communication is very low. Current standards operate at ~100 kHz, and indeed, lower frequencies of operation are likely to be acceptable for many applications. Thus, the digital portion of the circuit can potentially work in the realm of performance that is potentially achievable by printing. The two exceptions to the rule are the diode for rectification, and, in some applications, the clock generator.

The rectifier in an HF RFID application needs to be able to rectify at 13.56 MHz, and current standards require the clock be divided from this same frequency as well. The former appears to be possible with printed electronics. The latter is problematic; however it is possible to generate a kHz clock locally using an oscillator, i.e., without dividing down from 13.56 MHz. This introduces more variability and noise into the tag, but, for low data rates, the reader can screen this out. As a result, it is likely that a 13.56 MHz printed RFID tag is realizable, albeit with a local clock and a low data rate.[10,11] This appears to be usable for many applications including authentication, anti-counterfeiting, etc., and therefore, there is substantial industrial and research institute activity in this regard. The driver, of course, as in displays, is the development of printed transistors.

Printed Sensors

An area of recent interest for printed electronics is printed sensors.[12] The general concept is to develop simple sensors for product quality monitoring, etc., that use small numbers of transistors with very low performance. This is a diffuse area that is still developing, and therefore will not be discussed here. However, it is important to note that many of the concerns that apply to displays and RFID apply here as well, and therefore, the main technology driver for most areas of printed electronics continues to be the printed transistor.

TECHNOLOGY OVERVIEW

Printed electronics is a combination of a range of different fields, including semiconductor device physics and technology, ink chemistry and formulation, and printing technology. Given the strong interplay between the various areas, it is appropriate to consider printed electronics from all three perspectives in a unified treatment. To facilitate this, in this section, printed electronics will be considered using the following outline: First, printed transistors will be introduced, to identify the key materials and printing challenged. Based on this analysis, next, materials for printed transistors will be introduced, with the goal of identifying candidate inks for printed electronics. This will enable an analysis of the suitability of various printing techniques for printed electronics. Here, the focus will be on the advantages and disadvantages of inkjet printing for the same applications.

Printed Transistors: An Overview

As discussed above, the driver for many printed electronics applications is the printed transistor. The printed transistor is essentially a thin film transistor fabricated using printable materials. In other words, in its ultimate implementation, all three major material components of the printed transistor, i.e., the conductive electrodes, the insulating gate dielectric, and the semiconducting channel material, are printed.

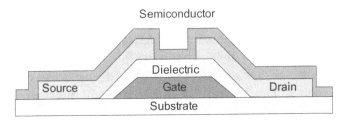

Fig. 6.   Cross-section of an archetypal printed transistor. The specific configuration shown is a bottom-gate configuration, with the gate below the channel.

Printing of transistors introduces several considerations that are substantially different from printing considerations in graphic arts. First, printed transistors are inherently three-dimensional devices with topographically rich surfaces. As a result, step coverage, wetting, pinholes, etc., are serious concerns in printed transistors. These issues will be considered in more detail later. Second, printed transistors require very careful alignment between layers. For example, in an ideal device, the edges of the gate should line up very closely with the edges of the source and drain electrodes, such that there is only a very small overlap between the gate and the source/drain electrodes. Excessive overlap leads to degraded switching speed due to overlap capacitance, while underlap leads to very poor performance due to extremely high series resistance. Given that overlap in printed electronics is almost always achieved using mechanical means (i.e., by controlling the position of the substrate relative to the layer being printed), this imposes severe constraints on achievable performance and accuracy requirements for mechanical parts within the print system. Third, printed transistors ideally use extremely thin, uniform layers, particularly for the gate dielectric and semiconductor, since these have tremendous impact on electrostatic integrity of the device.

From a materials perspective, the requirements vary by application, but in general, there is a trend towards low voltage operation and high carrier mobility. These two typically require the use of very well-ordered organic semiconductors with low defect density. Use of binders within the semiconductor ink is also typically not possible due to the degrading impact of these on carrier transport properties.

Overall, therefore, printed transistors impose numerous constraints on both materials and printing, which will be considered herein.

Printed Transistors: Process Integration

In general, printed transistors are realized in two common configurations — top-gate and bottom-gate. In the top-gate architecture, the channel is printed first, followed by the gate dielectric and gate electrode. In the bottom-gate architecture, the structure is flipped over, such that the gate and dielectric are printed before the channel. Each architecture has its own advantages and disadvantages, and the specific choice of architecture is strongly dependent on the choice of the material system.

Comparison of Top-gate and Bottom-gate
Transistor Architectures

In the top-gate architecture, the semiconductor is deposited before the gate dielectric and the gate electrode. This has several advantages. First, since the semiconductor is deposited on a known surface (the substrate itself, in typical examples), it is possible to exploit the fact that this substrate is typically extremely smooth and of known chemistry to ensure that the quality of the printed semiconductor is maximized. Most current in a transistor flows very close to the semiconductor-dielectric interface. In a top-gated transistor, therefore, this current flows near the top interface of the semiconductor; this may then be optimized to maximize the quality of the same. Additionally, since this layer is covered by the dielectric and gate, it may be protected from damage from subsequent process steps, etc.

The disadvantage of the top-gate architecture is that the semiconductor is the first deposited layer. As a consequence, this layer is potentially exposed to all the thermal cycles, solvent exposures, etc., of all subsequent steps, which may degrade the semiconductor.

Bottom-gated architectures, on the other hand, typically print the semiconductor as one of the last steps. This ensures that the semiconductor material is typically maintained in a high-quality state; the only subsequent layer will likely be an encapsulation step to protect the circuit. On the other hand, since the semiconductor in a bottom-gated device is printed after the gate and electrode, the surface on which it is printed is typically much rougher and less optimal than in a top-gated device; as a consequence, the ordering and morphology of the semiconductor film is often degraded relative to an optimized top-gated devices.

From an integration perspective, both devices therefore have advantages and disadvantages. The specific choice of architecture is impacted by the selected material system. For example, a bottom-gated architecture may be appropriate when using sintered contacts and a semiconductor that is thermally unstable, since the subsequent sintering steps required in a top-gated structure would degrade the semiconductor. Given this situation, therefore, it is appropriate to review both architectures individually, identifying the specific process steps as issues associated therewith.

Process Integration — Top-Gated Architecture

The process flow associated with a typical top-gated architecture is shown schematically in Fig. 7. From Fig. 7, some important process constraints are easily identified.

First, since the semiconductor is printed before the dielectric, it is crucial that the solvent used during the printing of the dielectric should not cause excessive degradation of the semiconductor film quality. Additionally, to realize pinhole-free dielectric films, the semiconductor layer itself should be very smooth. In particular, care must be taken at the edges near the contacts to ensure proper dielectric step coverage. Failure to do this may result in pinholes and thereby cause shorts between the gate and the source/drain electrodes.

Second, in the top-gated architecture, the typical process flow involves the semiconductor being printed before all other layers. This places important constraints on thermal processing; the semiconductor must be able to tolerate the temperature steps associated

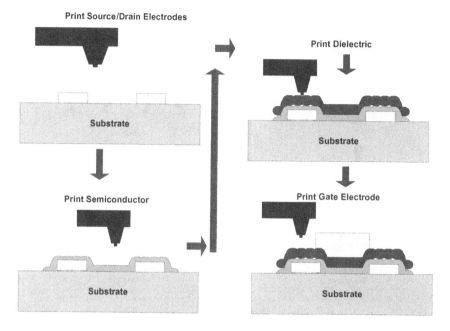

Fig. 7.   Conceptual process flow for fabricating a top-gated printed transistor.

with sintering all subsequent layers. This will limit somewhat the flexibility in choice of dielectric and conductor materials.

Process Integration — Bottom-Gated Architecture

The process flow used in a prototypical bottom-gated architecture is shown in Fig. 8.

In the prototypical bottom-gated architecture, the semiconductor is printed last. This offers maximum flexibility in terms of semi-conductor thermal and solvent compatibility, but places constraints on the choice of dielectric, etc. Clearly, in this device, just as in the top-gated architecture, dielectric pinholes between the gate and the source/drain are a critical concern. Since there is substantial topo-graphy in a bottom-gated architecture, step coverage of the dielectric is a key optimization parameter; should the dielectric "roll-off" the edge of the gate, a short will result, killing the device.

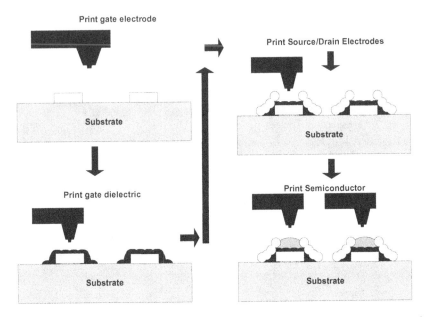

Fig. 8. Conceptual process flow used to fabricate a bottom-gated printed transistor.

Bottom-gated architectures have been used in conjunction with sintered nanoparticle electrodes and cross-linked dielectrics.[13] This attests to the advantages of the bottom-gated architecture with regard to process integration. On the other hand, recent reports suggest that the best semiconductor performance for many semiconductors is typically obtained using top-gated architectures due to the interface and morphology advantages of the top-gated architecture as discussed above.

Critical Process Integration Issues in Printed Transistor
Fabrication

At this point, it is appropriate to review the various process integration issues that must be considered during printed transistor fabrication. These issues in turn impact material selection, which is covered in a separate section.

    Layer roughness — Printed transistors are inherently multi-layer complex devices. Roughness of individual layers is therefore a key parameter. Roughness of lower layers directly impacts roughness, coverage, pinholes, etc., of upper layers. As a consequence, roughness is a critical process parameter. Given typical fluid flow issues, high-frequency roughness (i.e., sharp peaks, etc.) is typically a critical parameter, though global roughness is also important. In top-gated architectures, semiconductor roughness is extremely important, while in bottom-gated architectures, gate roughness is likely to be the most critical parameter.

Pinholes — The gate dielectric is typically considered to be the most critical component of a field-effect transistor. From a process integration and manufacturing perspective, this is certainly true in printed transistors as well; it is crucial that the gate dielectric have no pinholes, since this will lead to shorts and non-functional devices. Pinholes in turn depend strongly on roughness and wetting characteristics, etc., and therefore, substantial layer-to-layer optimization is required in this regard.

Step coverage — From the process flow schematics shown previously, it is apparent that printed transistors inherently have substantial topology within their cross-sectional structure. As a consequence, step coverage becomes an important parameter in process optimization. Given the large steps (typically several tens of nm or more) and the use of relatively thin subsequent layers, it is important that the layers cover each other adequately; liquids must be able to coat the vertical sidewalls of steps during a multilayer print process. This places constraints on fluid viscosity, evaporation rate, wetting, etc.

Overall, the above analysis allows us to identify critical process parameters impacting device yield and performance. Optimization and monitoring of these same parameters is the key to realizing circuits meeting performance needs while simultaneously delivering the process yield required to make printed electronics a reality.

Printed Transistors: Materials

As discussed previously, there is tremendous current activity in the development of printed electronic materials, with the result that several classes of printed materials have been developed and demonstrated. Since printable materials development is progressing at a very rapid rate at this time, this survey cannot serve as an encyclopedia of printable materials; rather it surveys the state-of-the-art and identifies general trends in materials development.

Printable Semiconductors

The most active area of research in printable electronic materials has been the general area of solution-processable (and therefore, potentially printable) semiconductors. In general, the materials development activity in printable semiconductors can be broken down into work on (1) soluble organic semiconductors, (2) soluble organic semiconductor precursors, and (3) soluble inorganic semiconductor precursors.

Soluble Organic Semiconductors

The vast majority of activity in printable semiconductors has focused on the development of soluble organic semiconductors, including polymers and soluble oligomers/small molecules. Generally speaking, activity in the area of polymers has focused on polythiophenes and polyarylamines,[14–16] since these have generally offered higher mobility than other polymer families. Over the last several years, mobility in these material systems has been steadily improving, with several reports of mobilities as high as 0.1 cm²/Vs having been reported. Indeed, most recently, there are reports of mobilities as high as 1 cm²/Vs having been made. Most polythiophenes and polyarylamines reported to date are p-type semiconductors, though n-type derivatives have been reported as well. This has been achieved by shifting the energy levels of the semiconductors.

One of the generally reported problems with organic semiconductors has been their air stability, though recent reports suggest that even the more unstable polymers can be moderately stable if used with appropriate contacts and dielectrics. In recent years, substantial

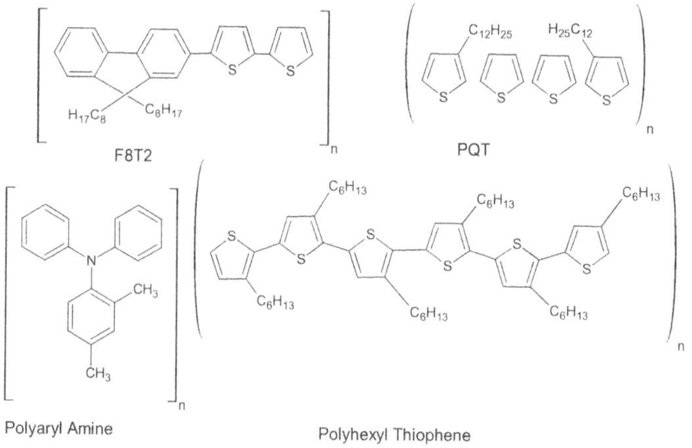

Fig. 9. Structure of some well-known printable polymer semiconductors.

progress has been made in synthetically modifying these polythiophenes to make them more air stable, and many of the more recent high-mobility polythiophenes show very good air stability. There are still some issues remain with these materials related to bias stress, however.[17] When transistors are fabricated with these materials, the performance shifts during continued use — this results in history-dependent switching behavior and degraded performance, which is a serious concern for circuit operation.

In addition to work on polymers, there have been several recent results on soluble oligomer semiconductors.[18] In particular, there has been work on various oligothiophene and acene derivatives. The main potential advantage of oligomers is that they tend to form

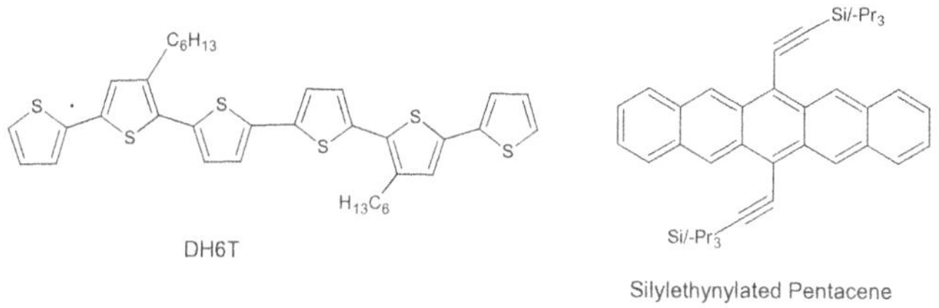

Fig. 10. Structure of some printable oligomer semiconductors.

strongly crystalline films, resulting in higher mobility. Certainly, this appears to be true at this time — the highest reported mobility for a solution-processed organic semiconductor has been achieved in a soluble acene derivative.

Soluble Organic Semiconductor Precursors

The materials discussed in the previous section are inherently soluble, and stay soluble even after they are incorporated into devices. This results in a possible integration concern. Printed transistors obviously require the printing of multiple layers of material over each other. If the materials are soluble, then there is a concern regarding solvent interactions between layers. An alternative approach that potentially simplifies integration is the use of convertible materials that are initially soluble and therefore printable. Upon printing, the resulting films are subjected to some form of energy (typically heat) to drive the conversion of the material to an insoluble form. During the conversion, the film typically reorders to form a crystalline film, resulting in transistors with high mobility.[19,20] Most work on this precursor-based route has focused on acenes and thiophenes. Both materials have been used to achieve mobility >0.1 cm^2/Vs, with reports of mobility as high as 0.8 cm^2/Vs having been achieved using a pentacene precursor. Molecular structures and conversion chemistry are shown in Fig. 11.

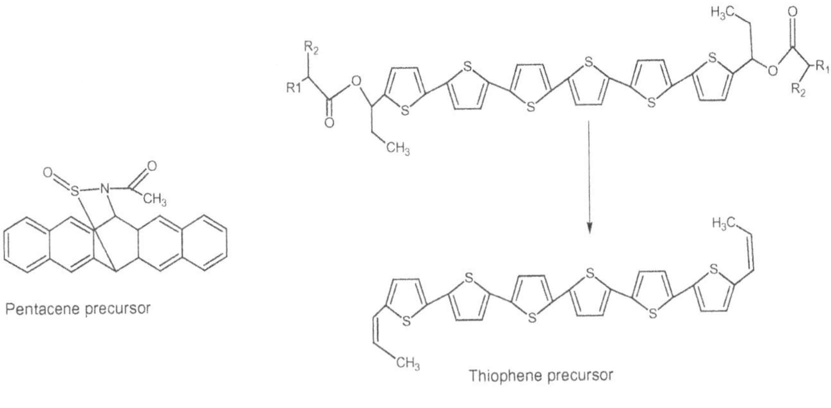

Pentacene precursor

Thiophene precursor

Fig. 11.   Some printable organic semiconductor thermally-convertible precursors.

Soluble Inorganic Semiconductor Precursors

In recent years, there have been several reports of printable inorganic material systems. These are particularly interesting since there is a clear possibility of using these materials to form highly stable, high performance printed films with performance exceeding that of organic materials. In general, several inorganic material systems have been studied in recent years, including various chalcogenides, zinc-oxide systems, and indeed, even silicon. Several routes to achieve stability have been studied. In contrast to organic materials, of course, all the reported inorganic semiconductors convert to insoluble form; in this sense, the printable materials are appropriately called semiconductor precursors. One of the earliest reported printable inorganic semiconductors was cadmium selenide. CdSe nanoparticles were synthesized and solubilized. After deposition and subsequent annealing, CdSe films result. There was not significant follow-up on this activity, and initial performance was poor.[21] However, it attests to the suitability of nanoparticles in this regard. More recently, zinc oxide (ZnO) nanoparticles have been used to realize solution-processed transistors, and mobilities greater than $0.1\,\mathrm{cm^2/Vs}$ have been reported.[22] Importantly, air stability does not appear to be a concern with this material system. Also, in recent years, there have been several other reports of printable inorganic semiconductors, including various ZnO-like compounds and several chalcogenides. Some reports made use of nanoparticle routes, while others made use of direct solution-based film formation techniques.[23,24] While the field of printable inorganic semiconductors is generally substantially less mature than that of organic semiconductors, progress is rapid.

Printable Dielectrics

Dielectrics are used as the gate dielectric in printed transistors. In this role, dielectrics are expected to provide good coupling between the gate electrode and the channel material, thus ensuring good electrostatic operation of the printed transistor. Typically, this is achieved by using as thin a dielectric as possible, to ensure maximized coupling between the gate and channel. Furthermore, this dielectric must provide good interfaces, particularly to the channel, and should be integratable into the printed device structure with minimal degradation of the layers above/below it.

Polymer Dielectrics

Just as polymers may be used to form printable semiconductors, so they may be used to form dielectrics as well. Indeed, polymer dielectrics are in widespread use in conventional microelectronics as well. For printed electronics applications, polymer dielectrics are therefore a natural choice for use in printed transistors. Several families of polymer dielectrics have been studied and used in printed transistors. These include various polyimides and other polymer dielectrics such as polyvinylphenol (PVP).[25] In general, these dielectrics are characterized by the following properties:

Solution processability — For cost reasons, it is desirable to have a printable dielectric. Therefore, as with the semiconductors discussed above, it is desirable to have solution processable dielectric materials. In particular, for dielectric applications, it is crucial that the solution processing technology be extremely robust and not result in any pinholes, cracks, etc., since these would form points through which electrical shorts might form.

Thermal compatibility — To increase electrical robustness, many commonly studied dielectric materials, including the polyimides and PVP above are annealed after printing. The purpose of this anneal is varied. In some materials, it serves to cause the evaporation of residual solvent. In other materials (for example, the polyimide and PVP above), it is used to cause a chemical conversion such as a cross-linking event. In every case, it is important that the requisite thermal process be compatible with the substrate and all layers that have already been printed at the time of annealing.

Frequency characteristics — Polymer dielectrics typically show signification frequency dependence in their electrical characteristics. For example, many polymers show significant roll-off in their dielectric constant at high frequencies due to dispersion effects. It is therefore important that the dielectric properties be matched to the application needs at the frequency of interest.

Interfacial quality — The dielectric interfaces have tremendous impact on device performance. High trap density at the dielectric-semiconductor interface can cause substantial device performance degradation. In addition, the interface itself may alter the morphology and packing structure of the semiconductor (for example, in a bottom-gated architecture, the semiconductor is deposited on top of the dielectric; as a consequence, the dielectric strongly affects the ordering of the semiconductor, since it forms a template for the organization of the initial layers of the semiconductor).

Relative dielectric constant — In general, most polymer dielectrics have relative dielectric constants in the range of 2.5 to 4. Recent work suggests that using a low dielectric constant material is optimal for achieving high mobility in organic transistors, though there is some disagreement in the universality of this relationship. One disadvantage of using low dielectric constant materials is that very thin dielectric layers may be needed to maximize coupling between the gate and the channel material; since this is difficult in printed processes, this poses a problem in relation to the reduction in the operating voltage of typical printed transistors.

High-k Dielectrics

As mentioned above, the relative dielectric constant of most polymer dielectrics is in the range of 2.5 to 4. To enable low-voltage operation, it may be desirable to increase coupling between the gate and the channel material. One convenient method of increasing coupling is to increase the dielectric constant of the semiconductor (Note that data in recent years suggests that this may degrade mobility of organic semiconductors if the semiconductor is placed directly in contact with the high-k material; however, it should be possible to use high-k materials to boost coupling provided a low-k interfacial layer is used between the semiconductor and the dielectric — this is still an issue under some debate). Several techniques to boost dielectric constant have been reported in the literature. These include:

Use of high-k nanoparticles — Analogous to the nanoparticle conductors discussed previously, it may be possible to print high-k dielectrics using nanoparticles. To date, there have been significant difficulties in producing such films with low leakage, since there tend to be substantial defects such as voids, cracks, etc., in such films.

Use of high-k nanoparticles in polymer dispersions — To solve the void and cracking issue, it is possible to disperse nanoparticles in a polymer matrix. The drawback is that this typically limits the actual dielectric constant boost to fractional amounts, since it is difficult to achieve significant mass loading of the nanoparticles within the polymer, resulting in a final film with only a small fraction of nanoparticle additive.

Use of organometallic precursors — Numerous high-k-forming metals, such as Zr (which forms ZrO_2), Hf (which forms HfO_2), etc., are available in organometallic form. These organometallics may potentially be printed and subsequently annealed/oxidized to form the relevant high-k material. While there have been several initial results in this regard, it has generally been difficult to realize low leakage films at low temperature. Due to the substantial volume change that occurs during the organometallic to metal oxide transition, resulting films are typically extremely porous. To eliminate this porosity, it is typically necessary to anneal the films at high temperatures, which are not plastic compatible and therefore not particularly desirable for all printed RFID applications.

Printable Conductors

Conductors are required in several areas in printed electronics. Conductors are used to form low-resistance interconnects, and antennae, as well as to form contact electrodes within transistors. Based on conductivity requirements, a range of conductors exist, ranging from flake inks (typically not inkjettable, and therefore not considered here), to nanoparticle inks and polymer conductors.

Particle-based Inks

In recent years, substantial materials expertise has been developed worldwide in the production of ultra-small particles, with diameters < 100 nm. These particles are often called "nanoparticles", for obvious reasons. An advantage of such small particles is that they are often extremely stable in colloidal suspensions, enabling their use in inks with high mass loading of particles. The particles themselves are produced using a variety of techniques. "Dry" techniques involve the use of sputtering or ablating small particles of a solid metallic target, and subsequently collecting the same and formulating an ink. "Wet" techniques involve the use of chemical reactions to produce nano-sized particulates, which are subsequently collected, purified, and formulated into an ink.

When the diameter of metallic nanoparticles is reduced well below 100 nm, an intriguing physical phenomenon manifests itself. As the diameter of a particle is reduced, the ratio of the particle's surface area to its volume increases (remember, the surface area depends on r^2, while the diameter depends on r^3). As a consequence, the net properties of the particle depend much more strongly on the surface properties than on the bulk properties. In general, the bonding of atoms on a surface typically are weaker than the bonding of energies within the bulk of a material. As a consequence, when particle diameters are depressed below 10 nm, many metallic nanoparticles show a dramatic reduction in melting point. For example, gold nanoparticles with diameters of approximately 2 nm have been found to "melt" at temperatures around 100°C, while bulk gold melts at temperatures of approximately 1000°C. This depression in melting point has a very important consequence for conductivity. Films formed using such small particles may be annealed at very low temperatures, causing the particles to "melt" and fuse together, at least locally. This results in substantially better contact between particles than would exist if particles are merely touching each other, resulting in the possibility of realizing films with conductivity substantially closer to bulk conductivity than is possible using larger particles or flakes. Indeed, conductivity as high as 30–70% of bulk conductivity has been reported.[26]

To produce particles smaller than 10 nm, "wet" processes are typically used. In a prototypical wet process, a precipitation reaction is used to cause nanoscale particulates to form in a liquid containing precursor material. As the particulates form, they are quickly encapsulated with an organic ligand to protect them and prevent them from fusing together or growing too large in solution. The resulting organic-encapsulated particles are collected, purified, and formulated into an ink.

Organometallic Precursors

Organometallic precursors are chemical compounds containing metallic elements chemically bonded to various organic functional groups. By appropriate design of organometallic molecules, it is possible to produce molecules which cleave and volatilize at low temperature; upon heating, the organic component of the molecule breaks off from the metal and evaporates, leaving behind a metallic film. In recent years, there have been several reports of organometallic precursors for metallic trace formation.[27] There are several issues, however, which affect the way in which these materials can be used in antenna formation. Since the actual volume fraction of metal within a typical organometallic precursor is extremely small (typically only a few percent), the final film that results upon sintering of an organometallic precursor is either extremely porous or has to undergo significant compaction. As a consequence, organometallic precursors are typically not appropriate for producing thick films of metal. However, these materials are typically fairly good at producing thin films.

Polymer Conductors

Polymer conductors generally have conductivity that is several orders of magnitude worse than metals; for example, using nanoparticles, sheet resistances as low as milli-ohms per square have been realized. Polymer conductors, on the other hand, have conductivities as high as tens to hundreds of kilo-ohms per square. Polymer conductors do have certain advantages. They can typically be deposited

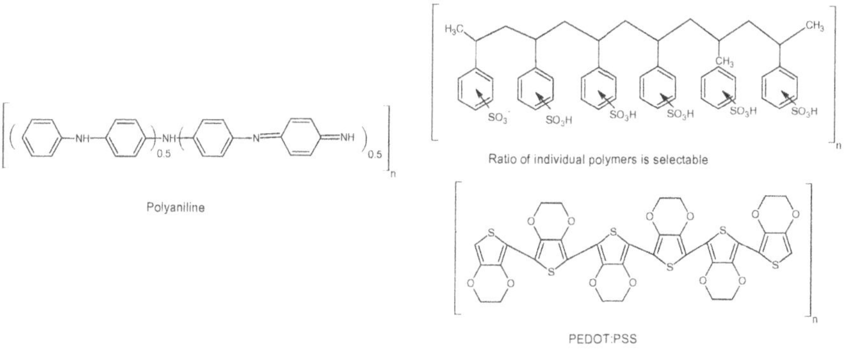

Ratio of individual polymers is selectable

Polyaniline

PEDOT:PSS

Fig. 12. Some commonly reported polymer conductors.

at room temperature and do not require a subsequent sintering step. Additionally, in many organic material systems, they form better interfaces to organic semiconductors than do metallic contacts. Therefore, there is some interest in polymer conductors as a means of realizing contacts in printed transistors.[28] Since the length of the conductor is small, the overall contribution of resistance is small enough that the use of polymer conductors may be possible without significant performance degradation.

Several polymer conductors are commercially available, and have been used in the demonstration of printed transistors. These include PEDOT:PSS, which is a commercially available polymer conductor, as well as various versions of polyaniline. The latter is typically doped with an acid or salt to increase conductivity. Both of these material systems are water soluble and easily printable. They also typically form good interfaces to organic semiconductors, making them attractive for use in printed transistors. As with polymer dielectrics, however, it is important to note that their usability with inorganic semiconductors is questionable, of course.

Printing Technology: Implications of Inkjet

Inkjet printing has been one of the most studied printing technologies for printed electronics. This is somewhat surprising, since inkjet printing as a whole is a relatively new and developing technology for high speed, low cost printing.

Advantages of Inkjet Printing for Printed Electronics

The reasons for the popularity of inkjet can be summarized as follows:

Compatibility — Inkjet printing allows the use of very low viscosity inks (1–20cP). This is a tremendous advantage of inkjet printing over more conventional analog printing techniques such as gravure printing, screen printing, etc. Many of the materials discussed above have somewhat limited solubility, which limits the achievable mass loading in stable ink formulations. Additionally, for many of the inks, the addition of binders is unacceptable, since these binders poison the electronic functionality of the ink. As a result, inkjet has tremendous advantages in this regard. For example, the previously discussed nanoparticle inks are easily formulated into nanoparticle inks using a range of organic solvents. The choice of solvent is driven by the need to have low evaporation rates (to prevent clogging in the head), good solubility (to maximize mass loading), and high viscosity (to improve jetting stability). Common solvents used for nanoparticles include alpha-terpineol, butylbenzene, anisole, etc. Similarly, organic semiconductor inks are often formulated in anisole and various toluene and benzene derivatives. Finally, some of the polymer conductors, such as PEDOT:PSS, are formulated in aqueous inks. As has been discussed in other chapters, multicomponent solvent blends are often used to allow careful optimization of film quality. Mass loadings are usually kept as high as possible (for example, typical nanoparticle inks are formulated with mass loadings between 5 and 30%, limited by the solubility of the particle). This ensures rapid film building with low porosity in the final film.

Digital input — Since inkjet allows for digital input, it allows for on-the-fly design changes. This is very important in research, since it allows for very rapid prototyping. Given the early stage of printed electronics, therefore, it is not surprising that inkjet has been so popular. A long term advantage of digital input technology is that it may allow for such operations as on-the-fly distortion correction, which may enable more accurate alignment over large area substrates.

Non-contact printing — Since inkjet does not use contact between the substrate and the printing head, it is relatively free from the main disadvantages of contact printing, namely degradation of the print quality over time due to abrasion of the print form, and also yield loss due to particles.

Resolution — Currently, it is possible to obtain commercial inkjet heads with resolution of 20 microns, and research mode heads have also been demonstrated with resolution better than 10 microns. In comparison, most analog print technologies produce features of worse than 30 microns (though some techniques have shown sub-10 micron resolution in research or controlled environments).

As a consequence of the above advantages, inkjet has received substantial attention as a means of realizing printed electronics. Unfortunately, to this point, inkjet has struggled to successfully make the transition from research to manufacturing. The reasons for this illustrate clearly the disadvantages and concerns with inkjet printing.

Challenges Associated with Deploying Inkjet
for Printed Electronics

Inkjet printing is a drop-by-drop technique. Patterns are built on the substrate through the assembly of individual drops. This has very important implications on printed electronics, and indeed, all the major concerns preventing successful commercialization of inkjet printing in printed electronics are related to the influence of the drop-based deposition. The major concerns with printed electronics based on inkjet are summarized below.

Line Roughness Concerns

As discussed previously, roughness is a significant concern for printed electronic devices in general and printed transistors in

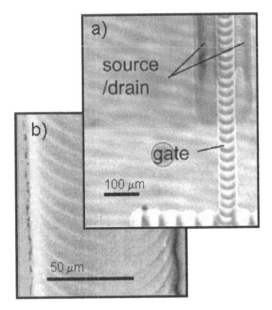

Fig. 13. Some sample inkjet printed lines from (a) Sirringhaus[29] and (b) Subramanian.[30] The pixilation-induced roughness in the lines is clearly visible.

particular. Roughness in lines degrades performance and can cause yield problems. For example, in a bottom-gated structure, a rough gate electrode can cause pinholes in the gate dielectric, making the transistor non-functional. To avoid this, researchers typically utilize thicker gate dielectrics. While this does allow for the realization of functional transistors, as discussed above, it does degrade operating characteristics of the device due to degraded gate coupling into the channel.

A survey of inkjet printed devices shown to date shows substantial roughness in the inkjet printed line, due to drops drying individually as the line is built. While functional devices are realizable, the resulting devices typically operate at high voltages due to the need for thick subsequent printed films.

The simplest solution to this problem is of course to allow drops to flow prior to drying; unfortunately, the disadvantage of this is degraded resolution since the flow that causes improved smoothness of the line surface also causes spreading of the line edges. By utilizing prepatterned surface energy treated substrates, this latter problem can be solved; however, this does add in another patterning step, and only works for initial printed layers.

The overall consequence, therefore, is that the operating window over which high resolution printing and smooth line surfaces are both achievable are convoluted and substantial engineering and study is required.

Drop Placement Accuracy Concerns

As discussed above, printed transistors depend on the alignment accuracy of the print system to achieve alignment. The drop-by-drop nature of inkjet printing has important consequences on alignment. Due to the variation in droplet ejection from the nozzle (caused by wetting, evaporative effects, etc.), a typical uncertainty in the solid angle of droplet impact on a surface will result in variability in drop placement accuracy of as much as 5 to 10 microns. This has two consequences. First, lines may meander as they are printed, setting a minimum gap between printed features in the same layer that is at least 2X the meander distance. Second, overlap to ensure good alignment must account for at least this much error in drop placement. This is clearly visible in the figure below, which shows the need for substantial overlap between the source/drain and gate in a printed transistor to account for meander in the drop placement.

This large overlap, of course, results in degraded performance of printed transistors. Given that the performance of printed transistors is marginal for many of the intended applications, this is a serious concern and is an area of intensive research.

Manufacturability Concerns

The final major concern with inkjet relates to the manufacturability of drop-by-drop pattern formation. For even relatively small circuits, the number of drops that must fire correctly very rapidly rises into the millions for many applications. This is a serious concern for printed electronics, since, unlike graphic arts, open circuits can be

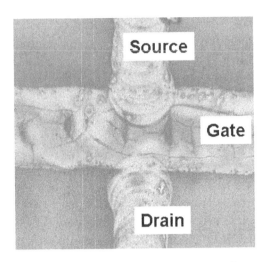

Fig. 14. Optical micrograph of a printed transistor showing large overlap between gate and source/drain. This overlap is necessary to account for large variability in drop placement accuracy.

an absolute yield killer. Currently, there is no tremendous amount of data available in this regard, but this is an area of emphasis which must be addressed before inkjet printed electronics can become a reality.

CONCLUDING REMARKS

Inkjet printed electronics is very attractive as a means of realizing potentially low cost circuits on flexible substrates. Potential applications range from displays to RFID tags to sensors. Over the last decade, a family of high-quality printable electronic materials has been developed, and processes for realizing printed devices have been demonstrated.

Inkjet printing has substantial advantages for printed electronics when compared to other printing techniques. However, important challenges do remain, and significant research and development is required to make inkjet printed electronics technically and economically feasible.

REFERENCES

1. Plummer JD, Deal M, Griffin PD. (2000) *Silicon VLSI Technology*, 1st ed. Prentice Hall.

2. Jang J. (2006) Displays develop a new flexibility. *Materials Today* 9(4): 46–52.

3. de Gans B-J, Duineveld PC, Schubert US. (2004) Inkjet printing of polymers: state of the art and future developments. *Adv Mater* 16(3): 203–213.

4. Bernius MT, Inbasekaran M, O'Brien J, Wu W. (2000) Progress with light-emitting polymers. *Adv Mater* 12(23): 1737–1750.

5. Chang S-C, Liu J, Bharathan J, Yang Y, Onohara J, Kido J. (1999) Multicolor organic light-emitting diodes processed by hybrid inkjet printing. *Adv Mater* 11(9): 734–737.

6. Comiskey B, Albert JD, Yoshizawa H, Jacobson J. (1998) An electrophoretic ink for all-printed reflective electronic displays. *Nature* 394: 253–255.

7. Burns SE *et al.* (2005) 12th Int Display Workshops/Asia Display, p. 16. Takamatsu, Japan.

8. Liang RC, Hou J, Zang HM. (2002) Microcup electrophoretic displays by roll-to-roll manufacturing processes. *Proc Int Disp Workshops* 9: 1337–1340.

9. Finkenzeller K. (2003) *RFID Handbook*. Wiley.

10. Myny K, Van Winckel S, Steudel S, Vicca S, De Jonge S, Beenhakkers M, Sele C, van Aerle N, Gelinck G, Genoe J, Heremans P. (2008) An inductively coupled 64b Organi RFID Tag operating at 13.56MHz with a data rate of 787b/s. *Int Solid State Circuits Conf*, paper 15.3.

11. Knobloch A. (2005) Printed RFID labels based on polymer electronics. *Digital Fabrication*: 121–123.

12. Chang JB, Liu V, Subramanian V, Sivula K, Luscombe C, Murphy AR, Liu J, Frechet JMJ. (2006) Printable polythiophene gas sensor array for low-cost electronic noses. *J Appl Phys* 100: 014506.

13. Subramanian V, Frechet JMJ, Chang PC, Huang DC, Lee JB, Molesa SE, Murphy AR, Redinger DR, Volkman SK. (2005) Progress towards development of all-printed RFID tags: Materials, processes, and devices. *Proc IEEE* 93: 1330–1338.

14. Bao Z, Lovinger AJ. (1999) Soluble regioregular polythiophene derivatives as semiconducting materials for field-effect transistors. *Chem Mater* **11**(9): 2607–2612.

15. Wu Y, Liu P, Ong BS, Srikumar T, Zhao N, Botton G, Zhu S. (2005) Controlled orientation of liquid-crystalline polythiophene semiconductors for high-performance organic thin-film transistors. *Appl Phys Lett* **86**: 142102.

16. Ong BS, Wu Y, Liu P, Gardner S. (2004) High-Performance semiconducting polythiophenes for organic thin-film transistors. *J Am Chem Soc* **126**(11): 3378–3379.

17. Chang JB, Subramanian V. (2006) Effect of active layer thickness on bias stress effect in pentacene thin-film transistors. *Appl Phys Lett* **88**: 233513.

18. Park SK, Jackson TN, Anthony JE, Mourey DA. (2007) High mobility solution processed 6,13-bis(triisopropyl-silylethynyl) pentacene organic thin film transistors. *Appl Phys Lett* **91**: 063514.

19. Afzali A, Dimitrakopoulos CD, Breen TL. (2002) High-performance, solution-processed organic thin film transistors from a novel pentacene precursor. *J Am Chem Soc* **124**(30): 8812–8813.

20. Murphy AR, Frechet JMJ, Chang PC, Lee JB, Subramanian V. (2004) Organic thin film transistors from a soluble oligothiophene derivative containing thermally removable solubilizing groups. *J Am Chem Soc* **126**: 1596.

21. Ridley BA, Nivi B, Jacobson J. (1999) All-inorganic field effect transistors fabricated by printing. *Science* **286**: 746.

22. Volkman SK, Molesa SE, Lee JB, Mattis BA, Vornbrock Adela F, Bakhishev T, Subramanian V. (2004) A novel transparent air-stable printable n-type semiconductor technology using ZnO nanoparticles. *IEEE International Electron Device Meeting Tech Digest* 2004, pp. 769.

23. Redinger D, Subramanian V. (2007) High-performance chemical-bath-deposited zinc oxide thin-film transistors. *IEEE Trans Electron Dev* **54**: 1301–1307.

24. Norris BJ, Anderson J, Wager JF, Keszler DA. (2003) Spin-coated zinc oxide transparent transistors. *J Phys D Appl Phys* **36**(20): L105–L107.

25. Halik M, Klauk H, Zschieschang U, Schmid G, Radlik W, Weber W. (2002) Polymer gate dielectrics and conducting-polymer contacts for high-performance organic thin-film transistors. *Adv Mater* **14**(23): 1717.

26. Huang D, Liao F, Molesa S, Redinger D, Subramanian V. (2003) Plastic-compatible low resistance printable gold nanoparticle conductors for flexible electronics. *J Electrochem Soc* **150**(7): G412–G417.

27. Cheong MH, Wagner S. (2000) Inkjet printed copper source/drain metallization for amorphous silicon thin-film transistors. *IEEE Electron Dev Lett* **21**(8): 384–386.

28. Sirringhaus H, Kawase T, Friend RH, Shimoda T, Inbasekaran M, Wu W, Woo EP. (2000) High-resolution inkjet printing of all-polymer transistor circuits. *Science* **290**(5499): 2123–2126.

29. Burns SE, Cain P, Mills J, Wang JZ, Sirringhaus H. (2003) Inkjet printing of polymer thin-film transistor circuits. *MRS Bulletin* **28**(11): 829–834.

30. Redinger D, Farshchi R, Subramanian V. (2004) An ink-jet-deposited passive component process for RFID. *IEEE Trans Electron Dev* **51**: 1978.

CHAPTER 16

Ceramic Inks

Stefan Güttler
Fraunhofer Institute Manufacturing Engineering and Automation
Nobelstraße 12, 70569 Stuttgart, Germany

Andreas Gier
Inomat GmbH, Bildstockerstraße 16, 66538 Neunkirchen, Germany

INTRODUCTION

Ceramic inks are mainly of interest for two kinds of applications. The first is the decoration of ceramic tiles and dinnerware; the second is the freeform fabrication of ceramic components, an application of 3D printing. There are many studies in the literature concerning both applications (e.g., Refs. 1–4). As for manufacturing applications of inkjet technology in general, the deposition of ceramic inks is a still growing field; an example is the coating of gas sensors with ceramic nanoparticles. In this chapter some emphasis is laid on decoration with ceramic pigments.

The market for value priced dinnerware and wall tiles strongly tends toward flexible production of small batches and even single items, for example to replace valuable dishes no longer for sale or to fabricate customized tiles. State-of-the-art decoration with ceramic pigments is screen printing, but this technique does not allow small batch production due to the high costs of the screens. Alternative digital printing technologies are inkjet and laser printing.

For the highly resolved deposition of ceramic particles, in principle both technologies are applicable. Due to their different operation principles, both techniques have advantages and disadvantages, but the laser printing approach will not be discussed here.

Several difficulties are connected with the composition and application of ceramic inks. The first is the rheology of ceramic suspensions. The viscosity of suspensions in general shows a shear dependency that depends on the volume fraction of solids, the particles' size and shape, and the interparticle forces. During the printing process high shear rates are exerted on the ink, so the shear dependent viscosity of an ink may strongly influence the inkjet pumping mechanism. We take a closer look at the rheology of suspensions in the following section. Secondly, due to the high density of ceramics, the chemical stability of ceramic inks is an important issue. The sedimentation of solids within the print head leads to clogging of the nozzles. A third problem is the high abrasiveness of ceramic particles due to their hardness. This causes severe wear on the nozzles and results in insufficient functional life of print heads in technical applications. We will also return to this issue.

In addition to the applicability of ceramic inks to inkjet printers, they have to fulfill several properties required by the specific application. Ceramic pigments for decoration may be quite large particles up to $10\,\mu$m and since the color effect is much weaker compared to organic pigments, a greater concentration of the pigment on the ceramic substrate is required. In a subsequent baking step, the ceramic pigments are sintered. This requires that the particles survive temperatures which typically range from 850 to 1150°C and the coating to have melting and thermal extension properties similar to those of the enamel glaze of the ceramic substrates, e.g., tiles.

Ceramic suspensions can also be deposited by dispensing or extrusion from a nozzle. While printing is a parallel process, i.e., many nozzles per color or substance are involved — say 128 or more, dispensing and extrusion processes usually use a single nozzle. The rheological properties of the ceramic suspensions are different from those of inkjet inks; they may be much more viscous, i.e., contain a much higher fraction of solids. Applications are the freeform fabrication of ceramic components where a high spatial resolution (in the range of μm) can be reached. A recent material deposition technique using ceramic inks is described in Ref. 5.

RHEOLOGY OF CERAMIC SUSPENSIONS

The rheology of suspensions generally differs from fluids as a result of the hydrodynamic forces acting on the particles. The following figure illustrates this behavior in a print head. Since flows in print heads are in the range of low Reynolds numbers (Re $\sim$ 1–10 during drop formation in an ink channel with 350 μm diameter), the velocity profile within (circular) capillaries is parabolic. This is indicated in Fig. 1.

At very low shear rates (i.e., flow velocities), particles in a chemically stable suspension approximately follow the layers of constant velocities, as indicated in Fig. 2. But at higher shear rates hydrodynamic forces drive particles out of layers of constant velocity. The competition between hydrodynamic forces that distort the microstructure of the suspension and drive particles together, and the Brownian motion and repulsive interparticle forces keeping particles apart, leads to a shear dependency of the viscosity of suspensions. These effects depend on the effective volume fraction of

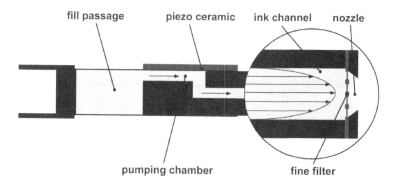

Fig. 1. Sketch of an inkjet print head. The parabolic velocity profile in the ink channel is drawn.

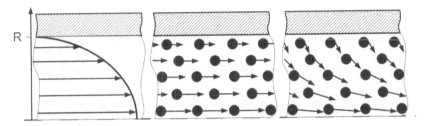

Fig. 2.   Laminar flow within a capillary: Hydrodynamic forces drive particles out of layers of constant velocity.

solids in the suspension, the particles' interaction (i.e., the strength of repulsion), and their size, shape and surface properties. The various physical effects which lead to shear thinning of suspensions at intermediate shear rates and (possibly) to shear thickening at high shear rates are described in the literature (e.g., Refs. 6–9). Here we restrict ourselves to a more qualitative understanding of these effects and the conclusions which are relevant for the inkjet printing of ceramic inks.

During the fire pulses high shear rates are exerted on the ink, about $\dot{\gamma} = 10^4 - 10^6 \cdot 1/s$, depending on whether single or multiple pulses are used to generate drops, i.e., if the print head is binary or grey scale. It is difficult to conduct experiments at such high shear rates, so to our knowledge all studies on the viscosity of suspensions have been obtained at lower shear rates.

An experiment with a dilute ceramic suspension was made as follows: A very small quantity of silicon carbide particles (d $\sim 6\,\mu$m) was dissolved in silicon oil ($\eta \sim 350\,$mPas). The suspension was pumped at high pressure through a glass capillary (d $= 0.6\,$mm). The experimental setup is shown in Fig. 3. The velocities of the silicon carbide particles in the capillary are detected by an optical sensor. From these data, the statistics of the particles' velocities is calculated. Due to the optical properties of the sensor, the particles are only detected in a wedge-like sector of the cross-section of the capillary. The measured velocity distribution of the particles (Fig. 4) depends on the shape of this sector and, additionally, on the measuring tolerances of the sensor.

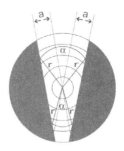

Fig. 3. Experimental set-up for medium to high shear rate experiments with a dilute ceramic suspension (left) and the sector of the cross-section of the capillary where particles are detected (right).

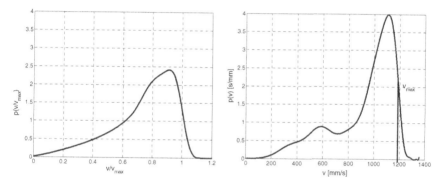

Fig. 4. Expected probability density for the particles' velocities if particles follow layers of constant velocities (left) and the probability density calculated from the measured velocities of the particles (right).

The suspension is highly accelerated (up to several $100\,\text{m}/\text{s}^2$) at the entrance of the capillary. Since the density of the particles is large, compared to the suspending fluid, the particles move relative to the fluid during the acceleration phase. The maximal shear rate at the border of the capillary is about $\dot{\gamma}_{\text{max}} = 8000\,1/s$. Under the assumption that the particles follow layers of constant velocities, as drawn in Fig. 2, the expected probability density of the particles' velocities is calculated. The result is shown in the left panel of Fig. 4. As mentioned above, the wedge-like sector of the cross-section,

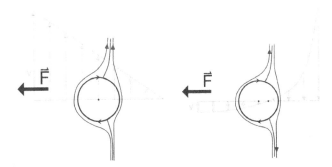

Fig. 5. Buoyant force acting on particles in a sheared flow (Magnus force). Left: Particle with velocity relative to the surrounding fluid. Right: Particle in a parabolic flow with no relative velocity.

where particles are detected, enters into this velocity distribution. For comparison, a typical measurement result is shown in the right panel.

As can clearly be followed, too many fast particles are detected because the particles are driven towards the center of the capillary by a buoyant force, the Magnus force.[10,11] Please note that the particles have not yet reached equilibrium. The origin of the Magnus force is sketched in Fig. 5 for the cases of a particle having a relative velocity to the surrounding fluid as it appears during the acceleration phase (and the fire pulses) and for a particle with no relative velocity. In the sheared flow the particles start to rotate with a thin boundary layer of fluid adherent to the surface. This causes the stagnation point of the surrounding flow to be shifted towards the wall of the capillary and consequently leads to a buoyant force directed to the center. The Magnus force decreases from the wall of the capillary towards the center where it vanishes; therefore, the particles are driven together as illustrated in Fig. 2. This is a qualitative point of view; in detail the Magnus force is intricate.[10,11]

The viscosity in the low shear regime depends mainly on the effective volume fraction of the particles in the suspension. There are many expressions given in the literature[12,13] which relate the low shear viscosity of a suspension η_0 to the viscosity of the suspending fluid η_s. Two formulas which are independent of parameters specific for a suspension are

$$\eta_0 = \eta_s(1 + 2.5\Phi + 5.2\Phi^2) \quad \text{for } \Phi < 0.1$$
$$\eta_0 = \eta_s(1 - \Phi/\Phi_m)^{-n} \quad \text{for } 0.1 < \Phi < \Phi_m$$

Φ denotes the effective volume fraction of solids and Φ_m the maximum effective volume fraction at which the flow of the suspension is blocked. In a chemically stabilized suspension, the volume of the particles is increased by the steric layers on their surface. This defines the *effective* volume of the particles, which depends on chemical stabilization by the suspension. By adding dispersants, the effective volume of the particles and therefore the effective volume fraction are increased, whereas the maximum packing density of particles decreases.[2] To obtain a quantitative formula, the maximum effective volume fraction of solid can be assumed $\Phi_m = 0.64$; this is the maximum volume fraction in a suspension of hard spheres, a model system.[9] Usually $n = 2$,[13] but n can be fitted to experimental data as well, e.g. Ref. 1. If one equates both formulas given above for $\Phi = 0.1$ and $\Phi_m = 0.64$, one obtains $n = 1.55$.

In the first formula only hydrodynamic contributions are considered, while in the second formula interparticle interactions are taken into account.[13] These formulas are expected to hold at least approximately. The decrease of the low shear viscosity of ceramic suspensions with increasing mean particle size at fixed solids concentration is reported in Ref. 1.

The shear dependency of the viscosity of ceramic suspensions at low to intermediate shear rates can be measured with common rheometers. The results for a ceramic suspension with a high volume fraction of solids and a ceramic ink are shown in Fig. 6.

The simplest model to describe shear thinning is the power model:

$$\eta(\dot{\gamma}) = \tilde{\eta}\dot{\gamma}^{(1-n)/n}$$

where $\dot{\gamma}$ denotes the shear rate and $\tilde{\eta}$ a constant containing the temperature dependency of the viscosity. For Newtonian liquids $n = 1$; $n > 1$ for shear thinning substances. For the suspensions shown in Fig. 6, $n \approx 1.8$ in the left panel and $n \approx 1.1$ in the right panel.

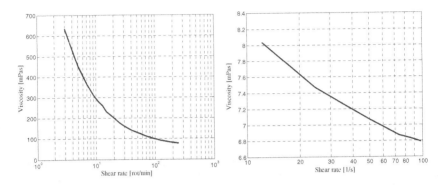

Fig. 6. Shear-dependency of the viscosity of a ceramic suspension with high volume fraction of TiO_2 and Ag nanoparticles (left) and of a ceramic ink containing about 15% solids by mass (right).

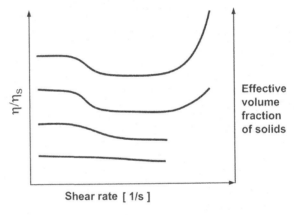

Fig. 7. Qualitative sketch of the shear dependency of the viscosity of stable suspensions over a large range of shear rates.

The qualitative behavior of the viscosity of suspensions over a large range of shear rates is depicted in Fig. 7. This type of shear dependency is found, for example, for concentrated colloidal suspensions.[9] These results cannot be immediately carried over to ceramic inks since the experiments are done with much higher concentrations of solids and at much lower shear rates than are applicable for the inkjet process. The suspended particles in these experiments are usually spherical at sizes of about 1 μm or smaller. But the shear dependency of the viscosity found for concentrated colloidal suspensions can be expected to also apply for ceramic inks, since theoretical works show that dilute colloidal suspensions have the same non-Newtonian flow behavior as concentrated suspensions.[8]

The viscosity and the shear dependency of the viscosity both increase with growing effective volume fraction of particles. For very low shear rates there is usually a low shear plateau η_0 and for high shear rates a high shear plateau η_∞ exists. These features are not described by the power model, but a more elaborate model such as the cross model is needed:

$$\eta(\dot{\gamma}) = \eta_\infty + \frac{\eta_0 - \eta_\infty}{1 + c\dot{\gamma}^m}$$

where c and m are fitting parameters (different from the parameters of the power model).

The most striking feature in Fig. 7 is the onset of shear thickening at some high shear rate. The instantaneous increase in viscosity of an ink during the fire pulse in fact would easily defeat the inkjet pumping mechanism. The rheological parameters that qualify a fluid to be applicable to inkjet printing are the viscosity and (dynamic) surface tension. Usual ranges of these parameters are $\eta = 6$ to 20 mPas and $\sigma = 25$ to 35 mN/m. Depending on the specific print head, these ranges may be larger or smaller. In the case of suspensions, the effective volume fraction of solids and the particle properties enter as additional parameters; i.e., if the volume fraction of solids of a chemically stable suspension is continuously increased where the low shear viscosity (and surface tension) are kept within the acceptable ranges, the inkjet process begins to get unstable at some point and finally fails. The stability of the inkjet process is regarded here as the probability of nozzle failure during operation. It is difficult to prove that shear thickening is the origin of this instability of the printing process, but results obtained for colloidal suspensions suggest that shear thickening may well occur at the high shear rates exerted during the fire pulse.

A qualitative understanding of shear thickening is that hydrodynamic forces (the Magnus force) drive particles close to physical contact such that hydrodynamic lubrication forces and frictional forces act between particles, leading to an increase in viscosity.[7,8] Shear thickening only appears if particles are permitted to approach sufficiently close. Therefore, whether shear thickening of a suspension occurs at high shear rates or a high shear plateau η_∞ is reached (Fig. 7), depends on the interparticle force, i.e., the strength of repulsion. Dispersants or surfactants added to inks may help to postpone or even suppress the onset of shear thickening.[8]

Shear thickening in colloidal suspensions is governed by particle size and surface roughness, in addition to the effective volume fraction of solids.[9] The onset of shear thickening depends on the effective particle radius, R; in the literature two relations for this dependency are found. The first is the Péclet number:

$$Pe = 6\pi\eta_s\dot{\gamma}R^3/kT$$

Pe should control the onset of shear thickening in colloidal suspensions. This being the case, the shear rate at which shear thickening occurs is (approximately) given by

$$\dot{\gamma}_c \sim 1/(\eta_s R^3)$$

In Ref. 7 it is argued that this relationship may hold for small values of Pe only, i.e., for small shear rates. An alternative approach is given which yields

$$\dot{\gamma}_c \sim 1/(\eta_s R^2)$$

If one finally considers that the Magnus force depends on the particles' radius as R^2,[10] the strong dependency of the onset of shear thickening on the particles' effective radius is obvious without concern for which relation exactly holds.

A different important issue in this context is the chemical stability of ceramic suspensions, which is indispensable for inkjet printing. For chemical stability, either steric or by charge, the surface to mass ratio is important: $S/m = 3/(\rho R)$. Due to the high density ρ of ceramics, a stable suspension also requires that the particles not be too large, i.e., be smaller than about 2 μm.

To summarize, problems with the application of ceramic suspensions with inkjet printers grow with the (effective) volume fraction of solids and the size of particles. Since the coloring effect of ceramic pigments is much weaker compared to organic pigments, a high volume fraction is wanted on the other side. The same applies for the freeform fabrication of ceramic components, where the layers should not be too thin. The trade-off is to use particles as small as possible.

Grinding ceramic pigments down to submicrometer sizes with a narrow distribution of the particles size is an expensive manufacturing step. Additionally, for many pigments the coloring effect improves with growing size; therefore the particles must not be too small to obtain bright colors. This especially applies to core shell pigments, see Fig. 9. Ceramic pigments used for decoration with inkjet printers are typically in the range of about 0.3–2 μm, depending on the pigment. This is already quite large since suspensions applied with inkjet printers usually contain particles clearly smaller than 1 μm. To prevent nozzle clogging, the maximum particle size should be less than about 1/10 of the nozzle's diameter. From this it follows that conventional print heads with nozzles sizes of 30–50 μm can be used.

CHEMICAL COMPOSITION OF CERAMIC INKS

The sol-gel method for coating substrates has been used so far in several industries: aerospace, optics, automotive, household machines and semiconductors. Sol-gel refers to a process in which a dispersion containing small molecules (the sol) forms a cross-linked polymer (the gel) upon drying. After the heating step, this xerogel is transferred into a dense film. This process is depicted in Fig. 8.[14,15]

The starting materials used in the preparation of the "sol" are usually inorganic metal salts or organic metal compounds. Metal

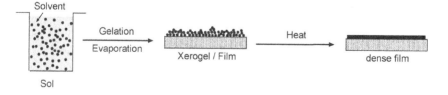

Fig. 8. Overview of the steps of the sol-gel process.

alkoxides and alkoxysilanes are the most popular precursors because they react readily with water. The reaction is called hydrolysis, because a hydroxyl ion becomes attached to the metal atom, as shown in the following reaction.

$$Si(OR)_4 + H_2O \leftrightarrow HO-Si(OR)_3 + ROH$$

The R represents a proton or other ligand such as an alkyl; so OR is an alkoxy group, and ROH an alcohol. Depending on the quantity of water and catalyst present, the hydrolysis may go to completion (i.e., all of the OR groups are replaced by OH) or it may stop when the alkoxysilane is only partially hydrolyzed. Two partially hydrolyzed molecules can link together in a condensation reaction, such as

$$(OR)_3Si-OH + HO-Si(OR)_3 \leftrightarrow (OR)_3Si-O-Si(OR)_3 + H_2O$$

or

$$(OR)_3Si-OR + HO-Si(OR)_3 \leftrightarrow (OR)_3Si-O-Si(OR)_3 + ROH$$

This kind of hydrolyses and condensation reaction leads to oligomer molecule structures. It should be noted that alcohol is not a simple solvent here. The hydrolysis and condensation reaction is reversible and the solvent is a reaction partner in this chemical equilibration.

To develop ceramic inks, hybrid organic-inorganic sols are first prepared by hydrolyzing methyltriethoxysilane and tetraethoxy-silane together with colloidal silica sol (10 nm in diameter). Ethanol is generated during the hydrolysis and condensation reaction by the silanes.[16,17]

To avoid later drying and filming of the ink in the nozzles of the print head, the ethanol is exchanged by distillation through hexanol or heptanol.

This coating material is the basis for incorporating ceramic coloring particles such as zircon red, cobalt blue or black. Zircon red is an inclusion pigment (core shell pigment) with an average size of about 5 μm. The cobalt blue and the black pigment are smaller; about 0.5–2 μm. REM pictures of the zircon red, the cobalt blue, and the black pigment are shown in Fig. 9.

Fig. 9. The core shell pigment zircon red (top) consists of 90–95% shell (SiO_2–ZrO_2) and 5–10% core (color). The cobalt blue pigment (middle) and the black pigment (bottom) are much smaller. (Note the different magnification.)

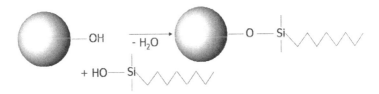

Fig. 10.   Scheme of surface functionalized coloring pigments by octylsilane.

The pigments are functionalized at the surface by ocyltriethoxysilane in order to stabilize these particles and to avoid agglomeration. This organic modification by the octyl group leads to a steric stabilization. A schematic draft is shown in Fig. 10.[18]

In the next step these particles are incorporated in the transparent silane matrix described above. In this network of silanes, the nano scaled SiO_2 particles on the one hand and the coloring pigments on the other can be dispersed without agglomeration. In Fig. 11 the chemical composition is shown schematically. To further improve chemical stability, the steric stabilization can be combined with an electrostatic stabilization. Depending on the incorporated pigments, these inks prove to be chemically stable up to several weeks.

The scheme of the chemical composition of the ceramic ink is visible in Fig. 11. Component A denotes the surface-modified coloring pigments, B the silane oligomers, C the SiO_2 nanoparticles, and D the silane monomers.

After printing the ceramic ink on ceramic substrates, the coatings are dried at about 150°C for 15 minutes. After the drying step the films are densified at about 1100°C for 15 minutes using a temperature profile which is compatible with the manufacturing process of the ceramic substrates.

The silane monomers and oligomers give a good chemical adhesion to the enamel glaze of the ceramic and also to the incorporated pigments. The high chemical and mechanical stability of the pure SiO_2 matrix is useful to yield high strength in the coated ceramic goods, e.g., tiles, in daily use.

It is known that the maximum thickness of sol-gel films is limited by the generation of tensile stress in the films during the densification

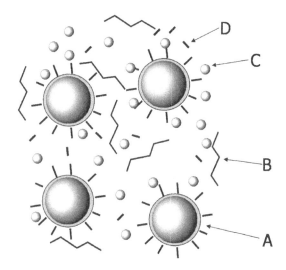

Fig. 11.   Draft of the chemical composition of the ceramic ink.

step. During the densification at high temperatures, the sol-gel coating shrinks as the pores in the xerogels collapse. According to the state of the art, sol-gel coatings have a thickness in the range of about 1 μm. Here the contained SiO_2 nano particles increase the solid content of the film and allow a maximum thickness of the coatings up to 10 μm.[19,20]

A different problem is the crack formation in the glaze during the densification step. At densification temperatures of about 1100°C the Tg of the enamel is reached and its viscosity decreases. The Tg of the coating is the same as for pure silica glasses, about 1400°C, since the coating consists of pure SiO_2. This leads to the formation of cracks in the glaze of the ceramics. To avoid this effect, a small quantity of alkali ions is added which acts as network modifier and decreases the Tg of the coating to about 1100°C.

DROP FORMATION AND STABILITY
OF THE INKJET PROCESS

For ceramic inks, the drop formation and stability of the printing process are studied as a function of the driving signal applied to the

Fig. 12.   An experimental inkjet printer developed at Fraunhofer IPA.

print head. An experimental printer used for these studies is shown in Fig. 12. In principle, piezo print heads all work similarly but the driving signals differ in detail. Therefore the parameters shown in Fig. 14 strictly apply only for print heads of the Dimatix SL-128 type (128 nozzles with a diameter of 50 μm). This print head works with a relatively strong viscous attenuation of the fire pulse. This leads to the drop formation being not very sensitive to the shape of the pulse signal, but mainly to the length and magnitude of the (trapezoidal-shaped) fire pulse. This is different from other types of print heads.

Photos of the drop formation for different values of the length and magnitude of the fire pulse are shown in Fig. 13. The time delay between the falling edge of the pulse and the flash of the stroboscope is constant in all pictures, so the distance of the drops to the nozzle is proportional to its velocity. The two drops on the right side will form undesired satellite drops, i.e., several drops instead of one, while the drop shown on most left is a bit slow. If the energy of the fire pulse is too low the printing process becomes unstable, even though the drop formation may look fine at some nozzles of the print head. As mentioned above, the stability of the printing process is regarded as the probability of nozzle failure during operation.

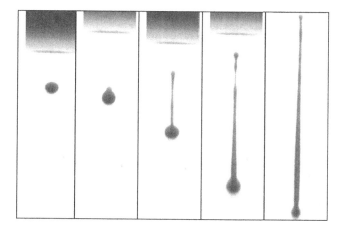

Fig. 13. Drop formation for different values of pulse length and magnitude for a ceramic ink.

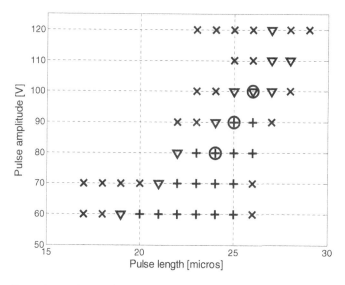

Fig. 14. Drop formation and stability of the inkjet printing process as a function of the driving pulse.

An example of the evaluation of the drop formation and stability of the printing process for a ceramic ink is shown in Fig. 14. Plus symbols indicate a good to acceptable drop formation, cross symbols stand for either the strong formation of satellite drops or the instability or failure of the printing process. Triangles indicate the border between both, since the evaluation is not always unique. One reason for this is the variation between different nozzles of a print head.

Stability tests are done for parameter settings which give at least acceptable drop formations. In this test about 1000 lines are repeatedly printed with increasing pauses in between. The failure of nozzles occurs when the printing process is started after a break. Single nozzles do not start at the first fire pulse but a few cycles later or fail completely. The parameter settings for which an acceptable printing reliability was obtained are indicated in Fig. 14 with a circle. Not all parameter settings for which a good drop formation is obtained allow a stable printing process. To a first approximation, the stability of the printing process increases with the energy of the fire pulse, i.e., with growing amplitude. If the pulse energy is too high on the other side, an ink film may form on the nozzle plate which impairs or prevents drop formation. The development of a film on the nozzle plate is prevented by the surface tension of the ink up to a limiting strength of the pressure wave.

Finally a test print on a tile is shown in Fig. 15. The pattern was printed with an ink containing 2.0% cobalt blue pigment and 8.0%

Fig. 15.   Cobalt blue pigment before (left) and after the baking step at 1100°C (right).

SiO_2 by mass. As can be seen clearly, the coloring effect decreases significantly after the baking step. The small cracks visible in the photo on the right are caused by the different melting temperatures and thermal expansions of the coating and the enamel glaze of the tile as mentioned above.

ABRASION OF NOZZLES

An important issue for any industrial application of inkjet printing of ceramic inks is the abrasion of the nozzles due to the hardness of the particles. An experiment demonstrating this is shown in Fig. 16.

The REM picture on the left side shows a nozzle of a print head used with fluidic inks only. To compare, on the right side a print head tested with ceramic particles is shown. The degradation of the nozzle made from tantalum occurred within 30 hours of use. The abrasion of the nozzles strongly increases with the size of the ceramic particles. This is also reported in Ref. 1 and gives another argument for keeping particles as small as possible.

One approach to prolonging the lifetime of the nozzles is to fabricate the nozzle plate from a (high performance) ceramic. Some types of the latest generation of print heads produced by Dimatix

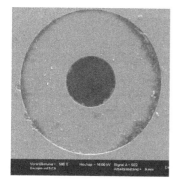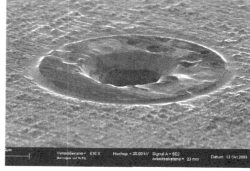

Fig. 16.   Used nozzles of a print head. The nozzle on the left was used with fluid inks only, the one on the right with an abrasive ceramic suspension for about 30 hours.

Com. are equipped with nozzle plates made from silicon. A possible alternative are print heads with nozzle plates made from plastic (polyimide) which are offered by Xaar Com. Flexible nozzle plates may cushion the impact of the ceramic particles. For the ceramic inkjet printers commercially offered today the problem of insufficient functional life of the print heads is not solved yet.

REFERENCES

1. Ainsley C, Derby B, Reis N. (2003) Viscosity and acoustic behaviour of ceramic suspensions optimized for phase-change ink-jet printing. *J Am Ceram Soc* **88**: 802–808.
2. Krishna Prasad PSR, Venumadhav RA, Rajesh PK, Ponnambalam P, Prakasan K. (2006) Studies on rheology of ceramic inks and spread of ink droplets for direct ceramic ink jet printing. *J Mater Process Tech* **176**: 222–229.
3. Obata S, Yokoyama H, Oishi T, Usui M, Sakurada O, Hashiba M. (2004) Preparation of aqueous pigment slurry for decorating whiteware by ink jet printing. *J Mater Sci* **39**: 2581–2584.
4. Zhao X, Evans J, Edirisinghe M, Song J. (2003) Formulation of a ceramic ink for a wide-array drop-on-demand ink-jet printer. *Ceramics International* **29**: 887–892.
5. Duoss EB, Twardowski M, Lewis J. (2007) Sol-gel inks for direct-write assembly of functional oxides. *Adv Mater* **19**: 4238–4243.
6. Bossis G, Brady JF. (1989) The rheology of Brownian suspensions. *J Chem Phys* **91**: 1866–1874.
7. Hoffman RL. (1997) Explanations for the cause of shear thickening in concentrated colloidal suspensions. *J Rheol* **42**: 111–123.
8. Bergenholtz J, Brady JF, Vicic M. (2002) The non-Newtonian rheology of dilute colloidal suspensions. *J Fluid Mech* **456**: 239–275.
9. Smith WE, Zukoski CF. (2004) Flow properties of hard structured particle suspensions. *J Rheol* **48**: 1375–1388.
10. Saffman P. (1965) The lift on a small sphere in a slow shear flow. *J Fluid Mech* **22**: 385.
11. Stone HA. (2000) Philip Saffman and viscous flow theory. *J Fluid Mech* **409**: 165.

12. Rutgers R. (1962) Relative viscosity and concentration. *Rheol Acta* **2**: 305–349.

13. Dávalos Orozco LA, del Castillo LF. (2006) Hydrodynamic behaviour of suspensions of polar particles. In Somasundaran P (ed.),*Encyclopedia of Surface and Colloid Science*, Vol. 4, pp. 2798–2820. Taylor & Francis Group, New York.

14. Brinker JC, Scherer GW. (1990) *Sol-Gel Science*. Academic Press, Boston, San Diego, New York, pp. 108–113.

15. Schmidt H, Kaiser A, Lentz A. (1986) *Science of Ceramic Chemical Processing*, pp. 87–93. John Wiley & Sons, Inc., New York.

16. Schmidt H, Scholze H, Kaiser A. (1984) Principles of hydrolysis and condensation reaction of alkoxysilanes. *J Non-Crystalline Solids* **63**: 1–11.

17. Fabes BD, Doyle WF, Zelinski BJJ, Silvermann LA, Uhlmann DR. (1986) Enhancement of fracture strength of cutted plate glass by the application of SiO_2 sol-gel coatings. *J Non-Crystalline Solids* **82**: 349–355.

18. Fabes BD, Berry GD. (1990) Infiltration of glass flaws by alkoxide coatings. *J Non-Crystalline Solids* **121**: 357–364.

19. Lange F. (1991) Microstructure, materials and applications. In Bradt RC (ed.), *Fracture Mechanics of Ceramic*, Vol. 2, pp. 599-609. Plenum, New York.

20. Mennig M, Jonschker G, Schmidt H. (1992) SPIE "Miniature and Micro-Optics" **1758**: 238–350.

Index

.

Printed in the United States
By Bookmasters